Interstate Liability for Climate Change-Related Damage

Interstate Liability for Climate Change-Related Damage

Elena Kosolapova

eleven
international publishing

Published, sold and distributed by Eleven International Publishing
P.O. Box 85576
2508 CG The Hague
The Netherlands
Tel.: +31 70 33 070 33
Fax: +31 70 33 070 30
e-mail: sales@budh.nl
www.elevenpub.com

Sold and distributed in USA and Canada
International Specialized Book Services
920 NE 58th Avenue, Suite 300
Portland, OR 97213-3786, USA
Tel: 1-800-944-6190 (toll-free)
Fax: +1-503-280-8832
orders@isbs.com
www.isbs.com

Eleven International Publishing is an imprint of Boom uitgevers Den Haag.

Cover Image: Ash Appleton

ISBN 978-94-6236-054-9
ISBN 978-94-6094-747-6 (E-book)
© 2013 Elena Kosolapova | Eleven International Publishing

Printed in The Netherland

Table of Contents

Foreword

In our current geological era, which has been aptly coined as the Anthropocene, the effects of human activities on the Earth's natural resources, including the atmosphere, are becoming visible. These effects include climate change resulting from human-induced emissions of greenhouse gases. Climate change creates opportunities as well as challenges for the human race. There will be winners and losers, but scientific evidence points to more losers than winners. In her monograph, Elena Kosolapova sets out to arm the losers with legal means to combat climate change. She identifies the legal opportunities for states, in particular least developed states and small island developing states, to challenge the business-as-usual attitude that many other states continue to assume. Although she recognizes that it may be difficult for states to obtain compensation for climate-change-related damage under international law, she convincingly demonstrates that states can invoke international law to force other states to reduce their emissions of greenhouse gases.

Mr. dr. R.J.M. Lefeber

Preface

If you have other things in your life – family, friends, good productive day work – these can interact with your writing and the sum will be all the richer.
David Brin

I would like to thank Malgosia Fitzmaurice, Frank Maes, Marc Pallemaerts, André Nollkaemper, and Rosa Uylenburg for taking the time to read and assess my work and for providing me with valuable comments on an earlier draft.

I thank René Lefeber for forcing me to think outside the box and helping me find a voice. Your inspirational guidance and continuous encouragement are greatly appreciated.

I would like to thank Marlon Boeve, Mireille van Rijn-Bogaart, Inge de Bruyne, Frank Groothuijse, Bibi Krot, Liselotte Smorenburg-van Middelkoop, Marjolein Visser, Liesbeth Vogelezang-Stoute, Josephine van Zeben, and especially Marleen Wessel, for their unceasing support and friendship.

A special word of gratitude goes to my mother, Natasha Nankman, for having faith in me, always.

And last, but not least, I thank my husband Chris du Plessis for his infinite patience, love, and generosity. Without you, it would not have been possible.

List of Abbreviations

AAUs	Assigned Amount Units
ADP	*Ad Hoc* Working Group on the Durban Platform for Enhanced Action
AFV	Alternative Fuel Vehicle
AR4	IPCC Fourth Assessment Report
ASEAN	Association of South East Asian Nations
AWG-KP	*Ad Hoc* Working Group on Further Commitments for Annex I Parties under the Kyoto Protocol
AWG-LCA	Ad hoc Working Group on Long-term Cooperative Action under the Convention
BAP	Bali Action Plan
BAT	Best Available Technology
CAA	Clean Air Act
CBD	Convention on Biological Diversity
CCS	Carbon Capture and Storage
CDM	Clean Development Mechanism
CERs	Certified Emission Reduction credits
CMEA	Council for Mutual Economic Assistance
CMP	Conference of the Parties serving as Meeting of the Parties to the Kyoto Protocol
COP	Conference of the Parties
CVP	Central Valley Project
EB	Enforcement Branch
EIA	Environmental Impact Assessment
EIS	Environmental Impact Statement
EIT	Country in Transition to a Market Economy
EPA	Environmental Protection Agency
ERT	Expert Review Team
ERUs	Emission Reduction Units
ESD	Ecologically Sustainable Development
FAO	Food and Agriculture Organization
FAR	IPCC First Assessment Report
FB	Facilitative Branch
GA	UN General Assembly
GATT	General Agreement on Tariffs and Trade
GCF	Green Climate Fund

GCRA	Global Change Research Act
GDP	Gross Domestic Product
GHG	Greenhouse Gas
IAEA	International Atomic Energy Agency
IBRD	International Bank for Reconstruction and Development
ICJ	International Court of Justice
ILC	International Law Commission
ILM	International Legal Materials
ICAO	International Civil Aviation Organization
IMO	International Maritime Organization
IMCO	Inter-Governmental Maritime Consultative Organization
IOM	International Organization for Migration
IOPC	International Oil Pollution Compensation
IPCC	Intergovernmental Panel on Climate Change
ITLOS	International Tribunal for the Law of the Sea
JISC	Joint Implementation Supervisory Committee
KP	Kyoto Protocol
LDC	Least Developed Country
LMO	Living Modified Organism
LULUCF	Land Use, Land Use Change and Forestry
MEA	Multilateral Environmental Agreement
MEPC	IMO Marine Environment Protection Committee
NAMA	Nationally Appropriate Mitigation Action
NAP	National Adaptation Plan
NEPA	National Environmental Policy Act 1969
NGO	Non-Governmental Organization
OCSLA	Outer Continental Shelf Lands Act
OJ	Official Journal of the European Union
OECD	Organisation for Economic Co-operation and Development
OPIC	Overseas Private Investment Corporation
PCA	Permanent Court of Arbitration
PCIJ	Permanent Court of International Justice
RMUs	Removal Units
SBI	Subsidiary Body for Implementation

SBSTA	Subsidiary Body for Scientific and Technological Advice
SC	UN Security Council
SDR	Special Drawing Right
SIDS	Small Island Developing States
SREX	IPCC Special Report on Managing the Risks of Extreme Events and Disasters to Advance Climate Change Adaptation
SWP	State Water Project
UK	United Kingdom of Great Britain and Northern Ireland
UN	United Nations
UNCC	United Nations Compensation Commission
UNCCD	United Nations Convention to Combat Desertification
UNCED	United Nations Conference on Environment and Development
UNCLOS	United Nations Convention for the Law of the Sea
UNCSD	United Nations Conference on Sustainable Development
UNECE	United Nations Economic Commission for Europe
UNEP	United Nations Environmental Programme
UNFCCC	United Nations Framework Convention on Climate Change
UNFF	United Nations Forum on Forests
UNRIAA	United Nations Reports of International Arbitral Awards
UNTS	United Nations Treaty Series
US	United States of America
USD	United States Dollars
VCAT	Victorian Civil and Administrative Tribunal
WMO	World Meteorological Organization
WTO	World Trade Organization
YILC	Yearbook of the International Law Commission

Introduction

In 1999, Kiribati's uninhabited islands Tebua Tarawa and Abanuea disappeared underwater. In 2007, the ocean engulfed a sandy islet in Micronesia's Pohnpei State and another islet split due to coastal erosion. Two more islands in the Chuuk Sea lagoon have now been completely submerged. Suparibhanga and Lochacharra on the Indian side of the vast Sunderbans, where the Ganges and Brahmaputra empty in the Bay of Bengal, have sunk and a dozen more nearby islands are under threat. A contemporary Inupiaq community of Shishmaref located on the island of Sarichef in Northwest Alaska has determined that the best way to adapt to the destructive beachfront erosion is to relocate the entire community to mainland Alaska. Devastating heat waves, wild fires, floods, droughts and hurricanes across the continents make the news on a continuous basis.

These reports conjure up fantastical images of the Plagues of Egypt and evoke the tale of Atlantis whose abysmal fate the inhabitants of island states and coastal communities now hope to avoid. The manifold manifestations of change affecting the climate system are expected to result in injurious consequences that will threaten sustainable habitation of the Earth.[1] Current and imminent damage from slow onset events (*e.g.* the rising sea levels, ocean acidification, desertification) and extreme weather events warrants a comprehensive response from the international community and local societies.

Climate change can be addressed locally through adaptation to its injurious consequences and internationally through concerted mitigation action. Largely based on mitigation and adaptation, today's international climate policy aims to prevent dangerous anthropogenic interference with the climate system.[2] Although mitigation and adaptation could significantly reduce the risks of climate change, they cannot avoid all climate change impacts,[3] and some degree of harm is unavoidable. Yet, the international climate regime does not address the injurious consequences of climate change. In 2010, the Conference of the Parties to the United Nations Framework Convention on Climate Change (also UNFCCC, or the Convention) established a work programme on approaches to address loss and damage associated with climate change impacts in developing countries that are

---

[1] See Chapter 1.
[2] 1992 UN Framework Convention on Climate Change, UNTS, vol. 1771, p. 107 (2000), Art. 2.
[3] IPCC *Fourth Assessment Report, Synthesis Report* (2007), p. 65.

particularly vulnerable to the adverse effects of climate change.[4] In 2012, parties to the Convention agreed to establish institutional arrangements, such as an international mechanism, to address loss and damage, but that will not happen before the nineteenth Conference of the Parties (COP) in 2013.[5]

In 2002, Tuvalu, a small island developing nation in the Pacific vulnerable to the rising sea levels, considered bringing a claim before the International Court of Justice (ICJ) against the United States (US) and Australia, major emitters in the industrialized world.[6] Possibly due to legal or political considerations, this plan never came to fruition. Instead, Tuvalu, together with other developing countries that are particularly vulnerable to the adverse effects of climate change, has been pursuing a multilateral negotiations strategy directed towards a comprehensive global agreement that would address climate-related risk management and risk reduction, including risk-sharing and risk-transfer mechanisms.[7] Risk management and transfer mechanisms, alongside with carbon dioxide capture and storage in geological formations (CCS),[8] may become effective adaptive strategies. However, many developing states do not have the economic capacities or the technological know-how to ensure timely preparation of their socio-economic infrastructure to the injurious effects of climate change. As more and more climate change impacts are felt around the world and given the slow pace of the international negotiations towards an effective international agreement, litigation may come as an attractive option, especially to states that are particularly vulnerable to climate change impacts.

The present work assesses the suitability of the law of state responsibility for climate change litigation. The research question can be formulated as follows: can injunctive relief, *i.e.* a court or arbitral tribunal order requiring a party to do – or to refrain from doing – something, provide an effective legal remedy for climate change-related damage within the framework of the law of state responsibility?

---

[4] *Cancun Agreements: Outcome of the work of the* Ad Hoc *Working Group on Long-term Cooperative Action under the Convention*, UNFCCC Decision 1/CP.16 (2010), para. 26.

[5] *Approaches to address loss and damage associated with climate change impacts in developing countries that are particularly vulnerable to the adverse effects of climate change to enhance adaptive capacity*, FCCC/CP/2012/L.4/Rev.1 (UNFCCC decision number not available at the time of writing), 8 December 2012, para. 9.

[6] 'Tuvalu threat,' ABC Local Radio, Australia, AM Archive, transcript from 4 March 2002, <www.abc.net.au/am/stories/s495507.htm> (last visited on 5 July 2012).

[7] See *e.g. Views on the work programme to consider approaches to address loss and damage*, submission of Grenada, on behalf of AOSIS, FCCC/SBI/2011/MISC.1 (28 February 2011).

[8] See *Modalities and procedures for carbon dioxide capture and storage in geological formations as clean development mechanism project activities*, UNFCCC Decision 10/CMP.7 (2011).

Chapter 1 sets the scientific background by describing the physics of climate change, including its causes, consequences, manifestations, and impacts on regions as well as on systems and sectors. It also addresses the main response strategies recommended by science, namely mitigation and adaptation. Chapter 2 sets the legal context by introducing the international climate regime governed by the 1992 United Nations Framework Convention on Climate Change, 1997 Kyoto Protocol (also KP, or the Protocol), and decisions taken by the COP and Conference of the Parties serving as Meeting of the Parties to the Kyoto Protocol (CMP). Chapter 3 identifies international liability regimes relevant to climate change-related damage. It distinguishes between state responsibility and state liability and outlines the various conceptual approaches to the latter based on: (1) the obligation to pay compensation; (2) the obligation to negotiate a redress settlement; (3) the obligation to ensure prompt, adequate, and effective compensation; and (4) the obligation to take response action. In the absence of interstate claims related to climate change, Chapter 4 turns to domestic litigation in order to identify some general principles of law which operate as barriers to a successful claim. It is argued that those legal principles could potentially cause obstacles to liability also at the interstate level. In dealing with liability for climate change-related damage in domestic courts, Chapter 4 analyses three groups of cases from a number of different jurisdictions: claims related to procedural injury; claims for compensation; and claims for injunctive and/or declaratory relief. Chapter 5 takes the analysis to an international level by applying the legal framework of the law of state responsibility to the climate change problem. It identifies the relevant primary obligations of states: international obligations on climate change mitigation; obligations on climate change adaptation; and the customary obligation to prevent significant transboundary harm. In determining the origins of state responsibility for breaches of the relevant obligations, four groups of states are distinguished in accordance with the principle of common but differentiated responsibilities and respective capabilities encapsulated in the UNFCCC:[9] industrialized states party to the KP; countries in transition to a market economy (EITs); industrialized states not party to the KP; and developing states. Chapter 5 then proceeds to address the content and the implementation of state responsibility for breaches of those obligations. Finally, Chapter 6 brings the analysis together by making some concluding remarks.

Focusing on liability, the present work does not address adaptive strategies related to CCS or risk assessment and transfer in respect of loss and damage associated with the adverse effects of climate change. From a methodological point of view, the research is largely descriptive insofar as it deals with climate change science

---

[9] 1992 UNFCCC, Art. 3(1).

(Chapter 1) and the international climate regime (Chapter 2). An analytical approach is adopted in Chapter 3 where the various international liability regimes are identified and the suitability of the existing conceptual approaches to liability to climate change-related damage is assessed. Chapter 4 is likewise analytical in its approach insofar as it seeks to explore some general principles of law emerging from climate change liability lawsuits litigated in domestic courts. In applying the legal framework of the law of state responsibility to breaches of state obligations related to climate change, Chapter 5 contains some normative elements.

1 Physics of Climate Change

1.1 Introduction

Climate change is arguably the greatest environmental challenge of our time. It has been the focus of scientific and academic research for several decades. Scientific findings on the link between human activities and the warming of the Earth's climate brought the issue to the top of the international political agenda. The international political response to the problem began in 1992 with the adoption of the UNFCCC, and its 1997 Kyoto Protocol committing industrialized nations and EITs to achieve quantified emissions reduction targets carried it to a new level. Yet, negotiations towards a comprehensive global mitigation framework are still continuing today.[10]

While the UNFCCC remains the primary forum to develop an international response to climate change, concerns about the changing climate have permeated many other international processes. Impacts on civil aviation and maritime sectors are being addressed by the International Civil Aviation Organization (ICAO) and International Maritime Organization (IMO) respectively.[11] Interlinkages between climate change and other environmental issues can be traced through the three Rio Conventions contributing to the same sustainable development goals, operating in the same ecosystems, and addressing interdependent issues.[12] The UN Conference on Sustainable Development (UNCSD) recognizes climate change as 'one of the greatest challenges of our time.'[13] Much of the work undertaken by the International Bank for Reconstruction and Development (IBRD) in developing countries, too, focuses on sustainable development and climate change[14] and the growing relevance of climate change-related issues to a number of other political processes cannot be underestimated.[15]

---

[10] See Chapter 2 for a discussion of the international legal regime governing climate change.

[11] *E.g.* for the ICAO's and IMO's respective climate change initiatives, see: <www.icao.int/environmental-protection/Pages/climate-change.aspx> and <www.imo.org/OurWork/Environment/PollutionPrevention/Air Pollution/Pages/GHG-Emissions.aspx> (last visited on 26 March 2012).

[12] 1992 UNFCCC, 1992 Convention on Biological Diversity, and 1992 UN Convention to Combat Desertification; sustainable development goals are reflected in Agenda 21, *i.e.* the Rio Declaration on Environment and Development and the Statement of Principles for the Sustainable Management of Forests.

[13] *The Future We Want: Rio +20 – UNCSD*, zero-draft of the outcome document (10 January 2012), p. 14.

[14] The World Bank climate change initiatives can be found at: <http://climatechange.worldbank.org /climatechange/> (last visited on 26 March 2012).

[15] *E.g.* 1971 Ramsar Convention on Wetlands, UN Forum on Forests (UNFF), Food and Agriculture Organization (FAO), International Organization for Migration (IOM), etc.

Already in the late 1980s, the UN General Assembly expressed concern that 'certain human activities could change global climate patterns, threatening present and future generations with potentially severe economic and social consequences' and requested immediate action leading to a comprehensive review and recommendations with respect to, *inter alia*, 'the state of knowledge and the science of climate and climatic change.'[16] The Intergovernmental Panel on Climate Change (IPCC), the leading intergovernmental body for the climate change science, was established in 1988 by the United Nations Environmental Programme (UNEP) and the World Meteorological Organization (WMO) to provide the international community with a clear scientific assessment of climate change and its environmental and socio-economic impacts.[17] Later that year, the UN General Assembly endorsed the action by the UNEP and WMO in a resolution on the protection of the global climate for present and future generations of mankind.[18] The role of the Panel, as defined by Principles Governing IPCC Work, is 'to assess on a comprehensive, objective, open and transparent basis the scientific, technical and socio-economic information relevant to understanding the scientific basis of risk of human-induced climate change, its potential impacts and options for adaptation and mitigation.'[19] To date, the IPCC has released four climate change assessment reports and numerous special reports, methodology reports, technical papers, and other supporting material. The most recent Fourth Assessment Report (AR4) was published in 2007 and the Fifth Assessment Report is scheduled for release in 2013-2014. The data contained in AR4 provides the most current scientific assessment of climate change and serves as the basis for this chapter.

Following its release, AR4 attracted some public criticism due to inaccuracies contained in the report. Since then, the IPCC was independently reviewed by the InterAcademy Council.[20] Following the InterAcademy Council recommendations, the IPCC undertook a complete review of its processes and procedures. Decisions on governance and management, conflict of interest, and procedures were adopted in June 2012.[21] Since the AR4 controversy, many scientists have reaffirmed the

---

[16] *Protection of global climate for present and future generations of mankind*, GA Res. 43/53 (6 December 1988).

[17] *Global Climate Change*, UNEP Governing Council Decision 14/20 (17 June 1987); WMO Executive Council Resolution 4 (1988).

[18] GA Res. 43/53 (6 December 1988).

[19] Principles Governing IPCC Work (1998), as amended in 2003, 2006 & 2012. Available from: <www.ipcc.ch/pdf/ipcc-principles/ipcc-principles.pdf> (last visited on 26 March 2012).

[20] *Climate Change Assessments*, review of the processes and procedures of the IPCC, InterAcademy Council (October 2010). Available from: <www.ipcc.ch/pdf/IAC_report/IAC%20Report.pdf>.

[21] The relevant decisions are available from: <www.ipcc.ch/organization/organization_review.shtml#.UOv--GdaHD4> (last visited on 8 January 2013).

validity of its key findings concluding that the failings in certain aspects of the report do not affect the overall scientific basis.[22]

The chapter briefly considers the science behind the legal issues examined in the present research. It begins its review of the causes and consequences of climate change by examining the role of anthropogenic (*i.e.* human) factors first. Second, the chapter describes the main consequences of the changing climate, *e.g.* global warming, and sea level rise and ocean acidification; and the ways those phenomena affect life on Earth. Third, it deals with climate change impacts across systems, regions, and sectors. Finally, the chapter shifts its focus to ways to address climate change, namely mitigation and adaptation, and closes with some concluding remarks.

1.2 Climate Change, Its Causes, and Consequences

Natural changes of the Earth's climate have played a fundamental role in the development of life as we know it. Basalt eruption events, sea-level falls, and asteroid impacts are among the possible causes of several major and lesser extinction events due to drastic environmental changes associated with those occurrences. Such changes include global cooling by enhanced cloud formation; global warming through the emission of carbon dioxide; and the production of acid rain and the destruction of the ozone layer attributed to nitrogen and sulphur compounds release into the atmosphere.[23] Solar variation, too, is hypothesized to have contributed to the Earth's surface temperature changes.[24] Natural factors have caused the climate to change in the past and will continue to do so in the future. Today, warming of the climate system is unequivocal.[25] Yet, natural environmental changes are not the sole cause of climate change. It has generally been accepted that human activities also contribute to it.[26]

---

[22] *Climate Change Assessments,* review of the processes and procedures of the IPCC, InterAcademy Council (October 2010), p. vii.
[23] N. MacLeod, Extinction, *in Encyclopaedia of Life Sciences,* London: Macmillan, 2001.
[24] J. Lean, J. Beer & R. Bradley, 'Reconstruction of solar irradiance since 1610: implications for climate change,' 22 *Geophysical Research Letters* 3195, 1995, at 3195.
[25] IPCC *AR4, Synthesis Report* (2007), pp. 30, 72.
[26] IPCC *AR4, Synthesis Report* (2007), p. 37.

1.2.1 Anthropogenic Factors

In 2007, the IPCC stated with 'very high confidence' that 'the global average net effect of human activities since 1750 [*i.e.* the beginning of the Industrial Revolution] has been one of warming.'[27] The Panel observed that global greenhouse gas (GHG) emissions due to human activities have increased by 70 per cent between 1970 and 2004 and further concluded that '[m]ost of the observed increase in global average temperatures since the mid-20th century is *very likely* due to the observed increase' in GHG concentrations of anthropogenic origin.[28] With respect to the past 50 years, the IPCC stated that the observed widespread warming of the atmosphere and ocean, together with ice mass loss, support the conclusion that it is 'very likely' that climate change 'is not due to known natural causes alone.'[29] It is 'very likely' that most of the global warming over the past 50 years is due to anthropogenic GHG increases and it is 'likely' that there is discernible anthropogenic warming 'averaged over each continent.'[30] The IPCC also detected 'discernible human influences [that] extend beyond average temperature to other aspects of climate, including temperature extremes and wind patterns,' and concluded that '[a]nthropogenic warming over the last three decades has *likely* had a discernible influence on observed changes in many physical and biological systems.'[31]

1.2.2 Global Warming

One of the main observed changes in the Earth's climate is the warming of the atmosphere. With the increase in GHG concentrations, more heat generated by the ultraviolet rays from the Sun is trapped in the upper layers of the atmosphere and prevented from going back into space. This causes the global temperatures to rise. The warming, in turn, entails other environmental shifts, such as the rise of the global average sea level due to thermal expansion of sea water and the melting of land ice in the mountains and Polar Regions, ocean acidification, changing weather patterns, and extreme weather events.[32] If business-as-usual behavioural strategies are maintained and the rates of GHG emissions are kept at current, or above current, levels, further warming of the atmosphere will 'induce many changes in the global climate system

---

[27] IPCC *AR4, Synthesis Report* (2007), p. 37.

[28] IPCC *AR4, Synthesis Report* (2007), p. 39, emphasis in the original; see also pp. 36, 72.

[29] IPCC *AR4, Synthesis Report* (2007), p. 39.

[30] Except Antarctica; IPCC *AR4, Synthesis Report* (2007), pp. 39, 72.

[31] IPCC *AR4, Synthesis Report* (2007), pp. 40- 41, emphasis in the original.

[32] On climate change and extreme weather events see: IPCC Special Report, *Managing the Risks of Extreme Events and Disasters to Advance Climate Change Adaptation* (SREX), Summary for Policymakers (2012), p. 5.

during [...] [this] century that would *very likely* be larger than those observed during the [...] [last] century.'[33] In the best-case scenario, the global temperatures will rise by 1.8°C with a likely range of 1.1-2.9°C by the end of this century; the worst-case scenario envisions a rise of 4°C with a likely range of 2.4-6.4°C.[34]

The IPCC has indicated with 'very high confidence' that physical and biological systems in terrestrial and marine ecosystems are already being affected by recent regional temperature increases.[35] There is 'high agreement' and 'much evidence' that with current climate change mitigation policies and related sustainable development practices, global GHG emissions will continue to grow over the next few decades.[36] And even if GHG concentrations are stabilized, anthropogenic warming and sea level rise will continue beyond the 21st century due to the time scale associated with climate processes.[37]

1.2.3 Sea Level Rise and Ocean Acidification

The global average sea level rise is driven by the rising temperatures and it is 'very likely' that anthropogenic emissions contributed to this process during the second half of the 20th century.[38] In combination with geological land changes, it is principally expected to affect coastal regions and small low-lying islands, some of which could disappear underwater because of an average sea level rise of merely several centimetres. Vast coastal territories could become uninhabitable due to the damaging effects of tidal surges, coastal erosion, and destruction of food crops. In 2009, the UN General Assembly expressed 'deep concern' that the 'adverse impacts of climate change, including sea level rise, could have possible security implications'[39] and jeopardize the very existence of some small low-lying countries.[40]

The average rate of the global average sea level rise over 1993 to 2003 was faster than that over 1961 to 2003 (about 3.1 millimetres per year as opposed to about 1.8), and the IPCC has stated with 'high confidence' that the rate of observed sea level rise increased from the 19th to the 20th century. It is estimated that in total over the last century the sea level rose by approximately 17 centimetres (12 to 22

---

[33] IPCC *AR4, Synthesis Report* (2007), p. 45.
[34] IPCC *AR4, Synthesis Report* (2007), p. 45.
[35] IPCC *AR4, Working Group II Report: Impacts, Adaptation and Vulnerability* (2007), p. 81; IPCC *AR4, Synthesis Report* (2007), p. 72.
[36] IPCC *AR4, Synthesis Report* (2007), pp. 44, 72.
[37] IPCC *AR4, Synthesis Report* (2007), p. 46.
[38] IPCC *AR4, Synthesis Report* (2007), p. 40.
[39] *Climate change and its possible security implications*, GA Res. 63/281 (11 June 2009).
[40] *Climate change and its possible security implications*, Report of the Secretary General, A/64/350 (11 September 2009), para. 71.

centimetres).[41] According to the IPCC, the end of this century will see a global average sea level rise of 18 centimetres in the best and 59 centimetres in the worst case scenario. These projections include 'increased ice flow from Greenland and Antarctica at the rates observed for 1993-2003;' they do not account for a possible further increase or decrease of Polar ice flow rates as it is uncertain how those may change in the future. If the contribution due to increased ice flow from the Polar Regions were to grow further with the global average temperature, the sea level rise could increase by an additional 10 to 20 centimetres.[42]

Another process associated with anthropogenic GHG emissions is that of ocean acidification by carbon dioxide. As the atmospheric concentrations of carbon dioxide grow, a greater amount of CO_2 is taken up by the ocean, which changes its chemical equilibrium. Once dissolved in ocean water, CO_2 forms a weak acid as pH levels decrease.[43] Anthropogenic carbon emissions since the Industrial Revolution have led to the ocean's acidification with an average decrease in pH of 0.1 units and a reduction in average global surface ocean pH of between 0.14 and 0.35 units is expected over the 21st century.[44] Although the effects of observed ocean acidification on the marine biosphere are undocumented, more acidic waters are expected to have detrimental impacts on marine shell-forming organisms, such as corals, as well as their dependent species.[45]

1.2.4 Desertification

On land, desertification is one of the greatest challenges to sustainable development linked to climate change. Desertification refers to 'land degradation in arid, semi-arid and dry sub-humid areas resulting from various factors, including climatic variations and human activities'[46] and affects millions of people worldwide. Dryland ecosystems cover more than a third of the world's land area. They are vulnerable to over-exploitation and inappropriate land use as well as to climatic changes. Although the relative importance of human and climatic factors in bringing about desertification is unclear, 'climate change and desertification remain inextricably linked because of feedbacks between land degradation and

---

[41] IPCC *AR4, Working Group I Report: The Physical Science Basis*, Summary for Policymakers (2007), pp. 6-7.
[42] IPCC *AR4, Synthesis Report* (2007), p. 45.
[43] IPCC *AR4, Working Group I Report: The Physical Science Basis* (2007), pp. 403-405.
[44] IPCC *AR4, Synthesis Report* (2007), p. 72.
[45] IPCC *AR4, Synthesis Report* (2007), p. 52.
[46] 1994 UN Convention to Combat Desertification, UNTS, vol. 1954, p. 3 (1999), Art. 1(a).

precipitation.'[47] This link is reflected in the close collaboration between the three Rio Conventions (see Section 1.1) and has also received ample recognition in the UN Convention to Combat Desertification (UNCCD) 10-Year Strategy. For instance, one of the outcomes under the policy operational objective reads: '[m]utually reinforcing measures among desertification/land degradation action programmes and biodiversity and climate change mitigation and adaptation are introduced or strengthened so as to enhance the impact of interventions.'[48]

1.2.5   Extreme Weather Events

In its recent special report on managing the risks of extreme weather events and disasters, the IPCC stated that a changing climate leads to changes in the frequency, intensity, spatial extent, duration, timing of, and can result in unprecedented, extreme weather and climate events.[49] There is 'high confidence' that exposure and vulnerability to extreme weather events will vary along spatial and temporal dimensions. It will also depend on various factors that determine the adaptive capacity of the affected area, *e.g.* economic, social, geographic, cultural, and environmental criteria.[50] According to the IPCC projections, the injurious effects of climate change are 'very likely' to increase as frequencies and intensities of some extreme weather events increase.[51] Significantly, there is evidence that 'some extremes have changed as a result of anthropogenic influences, including increases in atmospheric concentrations of greenhouse gases' albeit, admittedly, analysing and monitoring changes in extreme weather patterns is more difficult than calculating climatic averages.[52]

1.2.6   Climate Change Effects Across Systems, Regions, and Sectors

The effects of climate change stretch far and wide and many negative consequences are expected to come from the growing global average temperatures and rising sea levels. Yet, while all parts of the world are expected to sustain some form of climate change depending also on their respective vulnerabilities, the timing and magnitude of impacts relating to differing amounts and rates of climate change will vary across

---

[47] IPCC *AR4, Working Group II Report: Impacts, Adaptation and Vulnerability* (2007), p. 517.
[48] *The 10-year strategic plan and framework to enhance the implementation of the Convention (2008-2018)*, UNCCD Decision 3/COP.8, ICCD/COP(8)/16/Add.1 (2007), Ann., para. 11, Outcome 2.5.
[49] IPCC *SREX*, Summary for Policymakers (2012), p. 5.
[50] IPCC *SREX*, Summary for Policymakers (2012), p. 5.
[51] IPCC *AR4, Working Group II Report: Impacts, Adaptation and Vulnerability*, Summary for Policymakers (2007), p. 16; IPCC *AR4, Synthesis Report* (2007), p. 72.
[52] IPCC *SREX*, Summary for Policymakers (2012), p. 7; IPCC *AR4, Synthesis Report* (2007), p. 72.

regions. It is also 'likely' that some systems, regions, and sectors will be especially affected because of uneven distribution of climate change impacts and varying adaptive capacity. Climate change is projected to have especially damaging effects on ecosystems such as the tundra and mountainous regions, Mediterranean-type ecosystems and tropical rainforests, mangroves and salt marshes, coral reefs and the sea ice biome. Water resources 'in some dry regions at mid-latitudes [...], in the dry tropics and in areas dependent on snow and ice melt' will likely be put under serious strain by climate change.[53] Agriculture in low latitudes, low-lying coastal systems, and human health in populations with low adaptive capacity are expected to be among the worst affected sectors. Regions, such as the Arctic, Africa, small islands, and Asian and African megadeltas are expected to bear the worst of climate change impacts.[54]

Yet, not all climate change impacts are negative. For example, some health-related benefits are projected in temperate areas, *e.g.* fewer deaths from cold exposure, as well as some mixed effects, such as changes in malaria range and transmission potential in Africa.[55] Initial benefits are also projected for some regions in New Zealand by 2030.[56] Aggregate yields of rain-fed agriculture are expected in some North American regions in the first decades of this century.[57] Although in the Polar Regions specific ecosystems and habitats as well as indigenous ways of life are expected to be compromised, impacts on human communities resulting from the changing snow and ice conditions in the Arctic are projected to be of a mixed nature.[58] Additional opportunities are expected to result from adaptation and mitigation activities.[59] However, it is expected that initial benefits will be outweighed by the negative impacts of climate change.

1.2.6.1 Impacts on Systems and Sectors

Ecosystems: Overall, impacts of future climate changes on ecosystems are projected to place 20 to 30 per cent of plant and animal species at a 'risk of extinction if increases in global average temperature exceed 1.5 to 2.5°C,' and with a greater temperature increase, major changes in ecosystem structure and function, including negative consequences for biodiversity, are to be expected.[60]

---

[53] IPCC *AR4, Synthesis Report* (2007), p. 72.
[54] IPCC *AR4, Synthesis Report* (2007), pp. 48-52.
[55] IPCC *AR4, Synthesis Report* (2007), p. 48.
[56] IPCC *AR4, Synthesis Report* (2007), p. 50.
[57] IPCC *AR4 Synthesis Report* (2007), p. 52.
[58] IPCC *AR4, Synthesis Report* (2007), p. 52.
[59] IPCC *AR4, Synthesis Report* (2007), pp. 57, 60.
[60] IPCC *AR4, Working Group II Report: Impacts, Adaptation and Vulnerability*, Technical Summary (2007), pp. 59-63; IPCC *AR4, Synthesis Report* (2007), p. 48.

Food: While mean temperature increases of between 1 and 3°C are expected to increase crop productivity slightly at mid- to high latitudes (*e.g.* in North America and parts of Russia), temperature increases above that will decrease food production. At lower latitudes crop productivity is projected to decrease even with a temperature rise of 1 to 2°C.[61]

Coasts: Coasts are predicted to be exposed to increasing coastal erosion and sea level rise. By the end of the century, many millions more people than today are expected to experience annual floods due to the rising sea levels, with small island states and Asian and African megadeltas being affected the most.[62]

Societies: Communities that stand to be particularly exposed to the damaging effects of the changing climate are those whose livelihoods are closely connected with climate-sensitive resources and localities. Poorer communities are expected to be particularly vulnerable.[63]

Health: Climate change impacts on human health are expected to include: an increase in malnutrition due to reducing food availability; increased deaths and injuries associated with extreme weather events; increased frequency of cardio-respiratory diseases; and alteration in geographical distribution of infectious diseases. Benefits, such as reduced deaths from cold exposure, are expected to be outweighed by the negative impacts.[64]

Water: The melting of glaciers, mountain snow pack, and small ice caps as well as changes in precipitation are expected to reduce water availability in a number of regions thus posing a threat to the sustainable development of the affected regions.[65]

1.2.6.2 Impacts on Regions

Due to regional variability, climate change is expected to impact the world's regions differently: some will be affected worse and sooner than others. According to IPCC predictions, the worst effects will be endured in developing countries, partially due to their low adaptive capacity.

---

[61] IPCC *AR4, Working Group II Report: Impacts, Adaptation and Vulnerability* (2007), p. 285; IPCC *AR4, Synthesis Report* (2007), p. 48.

[62] IPCC *AR4, Working Group II Report: Impacts, Adaptation and Vulnerability* (2007), p. 327; IPCC *AR4, Synthesis Report* (2007), p. 48.

[63] IPCC *AR4, Synthesis Report* (2007), p. 48.

[64] IPCC *AR4, Working Group II Report: Impacts, Adaptation and Vulnerability* (2007), p. 418; IPCC *AR4, Synthesis Report* (2007), p. 48.

[65] IPCC *AR4, Working Group II Report: Impacts, Adaptation and Vulnerability* (2007), pp. 200-201; IPCC *AR4, Synthesis Report* (2007), p. 49.

Australia and New Zealand: Some ecologically rich sites in Australia and New Zealand are predicted to suffer significant biodiversity loss. The region is expected to be affected by intensified water security problems and reduced agricultural production by 2030. Coastal development and population growth in coastal areas will increase risks associated with sea level rise and the growing frequency and severity of storms and flooding.[66]

Europe: Magnified regional differences; increased risks of flash floods; more frequent coastal flooding and increased coastal erosion; frequent wildfires and heat waves on land are projected for the European region. Mountainous areas are expected to be affected with glacier retreat, reduced snow cover, and significant loss of species. The southern portion of the European continent is projected to experience increased drought conditions, reduced water availability, and decreased crop productivity.[67]

Latin America: By mid-century, Latin America is projected to undergo gradual replacement of tropical forest by savannah in eastern Amazonia, with associated risks of significant biodiversity loss, reduced agricultural productivity, and water availability stresses.[68]

North America: In the early decades of the century, North America is projected to see an increase in aggregate yields of rain-fed agriculture (with important regional variability) alongside with negative impacts such as winter flooding and reduced summer flows, stresses on coastal communities and habitats, intensification of heat waves, and exacerbated competition for over-allocated water resources.[69]

Asia: Some of the worst climate change effects are expected to affect Asia, particularly its megadeltas. Reduced freshwater availability, increased flooding, and pressures on natural resources associated with industrialization and economic development are among the impacts projected for the region.[70]

Africa: The African continent stands to be especially affected by climate change. Its vulnerability is exacerbated by its low adaptive capacity. Already by 2020, 75 to 250 million people in Africa are projected to be exposed to increased water stresses and in some African states agricultural production, including access

---

[66] IPCC *AR4, Working Group II Report: Impacts, Adaptation and Vulnerability* (2007), pp. 516-524; IPCC *AR4, Synthesis Report* (2007), p. 50.

[67] IPCC *AR4, Working Group II Report: Impacts, Adaptation and Vulnerability* (2007), pp. 549-558; IPCC *AR4, Synthesis Report* (2007), p. 50.

[68] IPCC *AR4, Working Group II Report: Impacts, Adaptation and Vulnerability* (2007), pp. 596-601; IPCC *AR4, Synthesis Report* (2007), p. 50.

[69] IPCC *AR4, Working Group II Report: Impacts, Adaptation and Vulnerability* (2007), pp. 627-635; IPCC *AR4, Synthesis Report* (2007), p. 52.

[70] IPCC *AR4, Working Group II Report: Impacts, Adaptation and Vulnerability* (2007), pp. 479-489; IPCC *AR4, Synthesis Report* (2007), p. 50.

to food, is projected to be severely compromised. It is expected that by the end of the century, heavily populated low-lying coastal areas will be affected by sea level rise, while the cost of adaptation could amount to 5-10 per cent of gross domestic product (GDP). By 2080, arid and semi-arid land is projected to increase by 5 to 8 per cent.[71]

Polar Regions: The Polar Regions, too, are identified among the most vulnerable to the changing climate. Reductions in 'thickness and extent of glaciers, ice sheets and sea ice [...] [together with] changes in natural ecosystems' are projected to have detrimental impacts on animal organisms and human populations.[72] However, some impacts on human communities in the Arctic, especially those resulting from changes in snow and ice conditions, are expected to be mixed.

Small Islands: Climate change is expected to place small islands under particular strain from the rising sea levels. Increased inundation, storm surge, erosion, and ocean acidification will affect local resources and the livelihoods of island communities. By mid-century, the growing pressure on water resources is expected to render them insufficient to meet demand during low-rainfall periods.[73]

Effectively, the fact that the largest share of historical and current GHG emissions has originated in developed countries[74] plays no part in the allocation of the injurious effects of climate change. It can be argued that countries that have contributed to climate change the least, *i.e.* developing states, particularly small island states and Africa, stand to suffer from it the most. The magnitude of regional impacts of climate change is exacerbated by developing states' low adaptive capacity.

1.3 Responding to Climate Change: Mitigation and Adaptation

Climate change can be countered in a number of ways at the local level as well as globally. Consistent with the IPCC recommendations, societies could adjust to its impacts (adaptation), reduce its rate and magnitude by way of GHG emissions reduction and limitation (mitigation) or these responses could complement each other. The capacity to adapt to and mitigate climate change is contingent on socio-economic circumstances, environmental conditions, and the availability of

---

[71] IPCC *AR4, Working Group II Report: Impacts, Adaptation and Vulnerability* (2007), pp. 444-451; IPCC *AR4, Synthesis Report* (2007), p. 50.
[72] IPCC *AR4, Working Group II Report: Impacts, Adaptation and Vulnerability* (2007), pp. 663-672; IPCC *AR4, Synthesis Report* (2007), p. 52.
[73] IPCC *AR4, Working Group II Report: Impacts, Adaptation and Vulnerability* (2007), pp. 695-703; IPCC *AR4, Synthesis Report* (2007), p. 52.
[74] See 1992 UNFCCC preamble.

information and technology.[75] The IPCC has proffered a number of mitigation and adaptation options albeit far less information is available on the costs and effectiveness of the latter.

Adaptation strategies can reduce societies' vulnerability to the injurious effects of climate change, both in the short and the long term. Adaptive capacity is intrinsically linked with social and economic development; however, it is not evenly distributed across and within societies.[76] It is subject to socio-political, economic, developmental, and cultural constraints. Even though developing countries are particularly susceptible to the injurious effects of climate change, societies with high adaptive capacity, too, can be vulnerable as evidenced by high mortality rates during the 2003 heat wave in Europe; the detrimental effects of Hurricane Katrina and Hurricane Sandy in the US in 2005 and 2012, respectively; and the destructive flash floods in Australia in 2012.

According to the IPCC, '[u]nmitigated climate change would, in the long term, be *likely* to exceed the capacity of natural [...] and human systems to adapt' to its injurious effects.[77] Delayed GHG emissions reduction 'significantly' constrains lower stabilization levels opportunities, thereby increasing the risk of more serious climate change impacts. The IPCC has stated with 'high confidence' that although neither mitigation nor adaptation alone can avoid all climate change impacts, together they could significantly reduce the risks of climate change.[78]

1.4 Concluding Remarks

Scientific findings set out in the IPCC assessment reports have motivated the UNFCCC process since its conception. Stabilization of GHG concentrations in the atmosphere at a level that would prevent dangerous anthropogenic interference with the climate system is the UNFCCC's ultimate objective and Parties have interpreted it to mean that the global average temperature rise must not exceed 2°C above preindustrial levels.[79] Compelling evidence of anthropogenic warming of the climate system presented in the IPCC AR4 can support informed decisions on the matter, including on increasing the level of ambition, in the context of the on-going negotiations, despite the absence of full scientific certainty. The following chapter describes the existing international legal framework for addressing climate change.

---

[75] IPCC *AR4, Synthesis Report* (2007), p. 56.

[76] IPCC *AR4, Synthesis Report* (2007), p. 56.

[77] IPCC *AR4, Working Group II Report: Impacts, Adaptation and Vulnerability* (2007), pp. 826-832; IPCC *AR4, Synthesis Report* (2007), pp. 65, 73.

[78] IPCC *AR4, Synthesis Report* (2007), p. 65.

[79] See 1992 UNFCCC, Art. 2; *Cancun Agreements: Outcome of the work of the* Ad Hoc *Working Group on Long-term Cooperative Action under the Convention*, UNFCCC Decision 1/CP.16 (2010), para. 4.

2 International Legal Regime Governing Climate Change

2.1 Introduction

The international climate regime is governed by rules, institutions, and procedures provided for in the 1992 UNFCCC, 1997 Kyoto Protocol, and decisions taken by the COP and CMP.[80] Today's international climate change policy largely rests on two pillars: mitigation of climate change and adaptation to its adverse effects. It is in this light that the present chapter describes the international climate regime. It analyses the international climate treaty framework by focusing first on the Convention and then on the Kyoto Protocol, including its market-based mechanisms and issues of compliance. Next, the chapter outlines the recent developments in the UNFCCC climate process noting the key highlights from the 2007 COP13/CMP3 in Bali to the 2012 COP18/CMP8 in Doha. The chapter concludes that although some degree of damage is unavoidable regardless of how successful international mitigation and local adaptation efforts are, the international climate regime is not (yet) equipped to address the injurious consequences of climate change. It is submitted that outside the climate change instruments, the law of state responsibility can provide a framework for ensuring that states take adequate mitigation measures even in the absence of explicit treaty provisions to that effect (see Chapter 5).

2.2 International Climate Regime

Twenty years after the first global environment conference in Stockholm and two years after the release of the IPCC First Assessment Report (FAR), the world governments gathered in Rio de Janeiro, Brazil, for the 1992 UN Conference on Environment and Development (UNCED), which later became known as the Earth Summit. It was attended by 172 states, 108 of which were represented by heads of state or government. Unprecedented in scale and magnitude, the Conference did not only lead to three major agreements on future approaches to sustainable development, namely Agenda 21 (a global action plan on sustainable development), the Rio Declaration on Environment and Development, and the Statement of Forest Principles. The Summit also produced two legally binding instruments that were

---

[80] See 1992 UNFCCC, Art. 7(2); 1997 Kyoto Protocol to the UNFCCC, UNTS, vol. 2303, p. 148 (2005), Art. 13(4).

opened for signature at the Conference: the Convention on Biological Diversity (CBD) and the UNFCCC.[81]

2.2.1   UNFCCC

The UNFCCC is widely considered the starting point of the international political response to the challenges of global climate change. It entered into force on 21 March 1994 and, having currently 195 parties (194 states and the European Union),[82] enjoys universal membership. Since then, eighteen meetings of the COP have been convened as well as multiple meetings of its two subsidiary bodies – the Subsidiary Body for Scientific and Technological Advice (SBSTA); and the Subsidiary Body for Implementation (SBI).[83] The Convention establishes a framework for action to achieve its ultimate objective of 'stabilization of greenhouse gas concentrations in the atmosphere at a level that would prevent dangerous anthropogenic interference with the climate system.'[84] This is a two-tier approach as it aims at (1) prevention of dangerous anthropogenic interference with the climate system, which, in turn, necessitates (2) stabilization of GHG concentrations in the Earth's atmosphere. Stabilization of GHG concentrations calls for mitigation measures. Yet, irrespective of mitigation efforts, some degree of climate change is unavoidable because of the past and current GHG emissions. Therefore, as explained in Chapter 1, societies need to adjust physical, biological, and human systems to the changes in the Earth's climate, *i.e.* take adaptation measures. It is important to note from the outset that even with the necessary mitigation and adaptation strategies in place, climate change will still lead to some injurious consequences. It is submitted that allocation of legal responsibility for climate change-related damage is imperative and will have broad implications for the international climate policy.

Guided by the need to protect the climate system for present and future generations, parties to the Convention agreed to be directed by the principle of common but differentiated responsibilities and respective capabilities whereby

---

[81] Negotiations on the UNCCD also began at the 1992 Earth Summit; the UNCCD was open for signature in 1994 and entered into force in 1996. The CBD, UNFCCC, and UNCCD became later known as the three Rio conventions.

[82] For current status of ratification see:
<http://unfccc.int/essential_background/convention/status_of_ratification/items/2631.php> (last visited on 3 April 2012).

[83] See 1992 UNFCCC, Arts. 7-10.

[84] 1992 UNFCCC, Art. 2.

developed nations take the lead in combating climate change.[85] This differentiation has its origin in the recognition of the fact that the largest share of historical GHG emissions has originated in developed countries and that whereas *per capita* emissions in the developing countries are relatively low, their share in the global emissions will grow to meet their social and development needs.[86] Thus, the developing countries' right to development and principles of equity have to be balanced against the global mitigation needs. With respect to mitigation of climate change, parties to the Convention have agreed to exercise precaution. The UNFCCC stipulates that in cases of threats of serious or irreversible damage lack of full scientific certainty should not be used as a reason for postponing measures that would 'anticipate, prevent or minimize the causes of climate change and mitigate its adverse effects.'[87] In this context, the precautionary principle offers the basis for 'taking action to minimize activities that could damage the environment even when the consequences of those activities are not fully understood.'[88]

Under the UNFCCC, developing and developed states alike are required to 'formulate, publish and regularly update national and, where appropriate, regional programmes containing measures to mitigate climate change.'[89] For developing countries, this obligation is contingent on the provision of finance and technology by developed countries included in Annex II.[90] In accordance with the principle of common but differentiated responsibilities and respective capabilities, only industrialized nations and EITs are placed under an obligation to '*adopt* policies and *take* corresponding measures on the mitigation of climate change' by limiting their GHG emissions and enhancing sinks (*i.e.* processes, activities or mechanisms that remove GHGs from the atmosphere) and reservoirs (*i.e.* components of the climate system, such as oceans, soils, and forests, that have the capacity to store, accumulate or release GHGs).[91]

---

[85] 1992 UNFCCC, Art. 3(1); for the purposes of the present analysis, the concept of equity is presumed to be included in the principle of common but differentiated responsibilities and respective capabilities.

[86] 1992 UNFCCC, preamble; see also Art. 3(4) reflecting the principle of sustainable development.

[87] 1992 UNFCCC, Art. 3(3).

[88] P. Cullet, 'Liability and redress for human-induced global warming: towards an international regime,' 43A *Stan. J. Int'l L.* 99, 2007, at 105.

[89] 1992 UNFCCC, Art. 4(1)(b).

[90] 1992 UNFCCC, Arts. 4(3), 4(7).

[91] 1992 UNFCCC, Art. 4(2)(a), emphasis added; see also IPCC *AR4, Working Group III Report: Mitigation of Climate Change* (2007), p. 820. In the present work, the terms 'industrialized country' and 'developed country' are used interchangeably to indicate the same group of states. In the UNFCCC context, those terms are used to refer to Annex I parties and in the KP context, they designate Annex B parties.

Specifically, industrialized countries agreed to aim at the common quantified target of returning their anthropogenic emissions of CO_2 and other GHGs to 1990 levels by the end of the 20th century.[92]

2.2.2 Kyoto Protocol

Albeit jointly met, the GHG emissions target set by the Convention proved to be insufficient to counter climate change. In fact, already in 1995, the need for revised emissions reduction targets was recognized by COP1 held in Berlin, Germany.[93] The Berlin Mandate started the process towards strengthening the commitments for developed countries. The Mandate placed the principle of common but differentiated responsibilities and respective capabilities at its core stating that it will 'not introduce any new commitments' for developing states.[94] At COP3 held in Kyoto, Japan, in 1997, the Berlin Mandate process culminated in a new binding legal instrument.

The Kyoto Protocol commits industrialized nations to reduce their combined GHG emissions by an average of 5.2 per cent compared to 1990 levels in the period 2008-2012 (the first commitment period) with specific emissions limitation and reduction targets varying per country.[95] Under the Protocol, EITs may use a base year other than 1990 (see also Section 5.3.4).[96] In 2012, CMP8 amended the Protocol agreeing on the second commitment period starting in 2013 and ending in 2020.

The Protocol provides for individual quantified GHG emissions limitation and reduction targets for developed countries.[97] In accordance with the principle of common but differentiated responsibilities and respective capabilities, developing countries are exempt from such targets. Developed countries are placed under a legal obligation to meet their individual emissions limitation and reduction targets in the 2008-2012 commitment period and, since the Doha amendment adopted in 2012, in the 2013-2020 commitment period. First and foremost, industrialized countries are required to achieve their targets by adopting and implementing domestic policies and measures towards reducing emissions by sources of GHGs

---

[92] 1992 UNFCCC, Art. 4(2)(b).

[93] *Berlin Mandate: Review of the adequacy of Article 4, paragraph 2(a) and (b), of the Convention, including proposals related to a protocol and decisions on follow-up*, UNFCCC Decision 1/CP.1 (1995).

[94] *Berlin Mandate: Review of the adequacy of Article 4, paragraph 2(a) and (b), of the Convention, including proposals related to a protocol and decisions on follow-up*, UNFCCC Decision 1/CP.1 (1995), para. 2(b).

[95] 1997 KP, Art. 3(1), Ann. B.

[96] 1997 KP, Art. 3(5), 3(6).

[97] 1997 KP, Art. 3(1), Ann. B.

and protecting and enhancing sinks for their removal.[98] However, they may also make use of the flexible market-based mechanisms to meet their reduction goals (see Section 2.2.2.1 below). Also, Parties to the Protocol agreed to pursue limitation or reduction of GHG emissions from international aviation and marine bunker fuels by working through the ICAO and IMO, respectively. Currently only emissions from domestic bunker fuels used for aviation and maritime transport are subject to emissions limitation and reduction commitments under the international climate regime.

During COP6 (Part 1) held in The Hague, the Netherlands, in 2000, signatories to the Protocol sought to elaborate its provisions; however, agreement could not be reached on a number of issues. Preceded by the US's announcement that it would not become party to the KP, COP6 Part 2 delivered a political package known as the Bonn Agreement. The detailed rules for its implementation were only adopted by COP7 in 2001 and are known as the Marrakesh Accords. The Kyoto Protocol entered into force in 2005 and currently has 192 parties (191 states and the European Union).[99]

2.2.2.1 Market-Based Mechanisms

Under the Kyoto Protocol, industrialized countries enjoy a certain degree of flexibility in meeting their emissions limitation and reduction targets. Obligations relating to domestic action against climate change are supplemented by the three market-based mechanisms: (1) Joint Implementation;[100] (2) Clean Development Mechanism (CDM);[101] and (3) International Emissions Rights Trading.[102] Each of the flexible mechanisms has its own governance procedures.

Joint Implementation and CDM are project-based mechanisms. The Joint Implementation mechanism allows industrialized countries to attain their emissions reduction targets under the KP by way of earning emission reduction units (ERUs) from an emissions reduction or emissions removal project in other developed countries (usually EITs). Joint Implementation provides a cost-effective way of

---

[98] 1997 KP, Art. 2(1)(a).
[99] For current status of ratification see:
<http://unfccc.int/kyoto_protocol/status_of_ratification/items/2613.php> (last visited on 4 April 2012).
[100] 1997 KP, Art. 6.
[101] 1997 KP, Art. 12.
[102] 1997 KP, Art. 17.

contributing towards the achievement of global targets since in some industrialized countries the abatement of GHGs is cheaper than in others.[103]

The CDM provides developed countries with an opportunity to invest in sustainable development projects that reduce emissions in developing countries with a view to earning certified emission reduction credits (CERs) that can be used towards meeting KP targets for the first commitment period. In addition to providing flexibility for industrialized countries, the CDM aims to promote sustainable development while at the same time reducing GHG emissions in developing states.

International Emissions Rights Trading enables industrialized countries to trade in GHG emission rights, which is why the process is commonly referred to as the carbon market. Under the KP, developed states with emissions reduction commitments have accepted individual quantified targets expressed as assigned amounts for the 2008-2012 commitment period, *i.e.* allowed levels of GHG emissions.[104] Those assigned amounts are broken down into assigned amount units, or AAUs, and developed states that have not used up all of their AAUs, can sell them on the carbon market to those industrialized countries that have exceeded their assigned amounts under the Protocol. There are units other than AAUs that can also be traded, namely: ERUs generated by Joint Implementation projects; CERs generated by CDM projects; and RMUs – removal units generated on the basis of land use, land use change and forestry (LULUCF) activities, *e.g.* reforestation.

The UNFCCC Secretariat monitors industrialized countries' emissions and keeps records of transactions carried out on the carbon market by keeping an independent transaction log called the International Transaction Log.

Participation in the KP carbon-market mechanisms is not unconditional. In addition to being necessarily supplemental to domestic action, they are subject to a number of eligibility requirements that developed countries must meet in order to avail themselves of their benefits. In order to make use of the carbon-market mechanisms, industrialized countries, *inter alia,* must: have ratified the Kyoto Protocol; have calculated their assigned amount in tonnes of CO_2 equivalent emissions; have in place a national system for estimating GHG emissions and removals in their territory; have in place a national registry to track the calculation and transfer of AAUs, ERUs, CERs, and RMUs and annually report this

---

[103] D.M. Ong, International Legal Efforts to Address Human-Induced Global Climate Change, *in* M. Fitzmaurice, D.M. Ong & P. Merkouris (Eds.), *Research Handbook on International Environmental Law* (450-470), Cheltenham: Edward Elgar, 2010, at 456.

[104] The use of market-based mechanisms in the second commitment period is described in Section 2.2.3.

information to the Secretariat; as well as annually report to the Secretariat information on GHG emissions and removals.[105]

In order to be entitled to participate in Joint Implementation and CDM projects, industrialized countries must meet further requirements of additionality, verification or validation, and reporting and review. First, both Joint Implementation and CDM projects must result in reductions 'additional to any that would otherwise occur,' *i.e.* emissions reduction, or an enhancement of removals by sinks, from those projects must be greater than that occurring in the absence of the certified project activity.[106] Second, Joint Implementation and CDM projects are subject, respectively, to verification and validation. Whereas 'Track 1' Joint Implementation projects (*i.e.* when a host Party meets all of the eligibility requirements) are verified by the host Party itself, verification of 'Track 2' Joint Implementation Projects (*i.e.* where a host Party does not meet all of the eligibility requirements) is carried out by an independent entity supervised by the Joint Implementation Supervisory Committee (JISC) and validation of CDM projects is performed by Designated Operational Entities using methodologies approved by the CDM Executive Board. And third, the KP market-based mechanisms are subject to the Protocol's reporting and review procedures under Articles 5, 7, and 8, the Marrakesh Accords, and the relevant CMP decisions.[107]

The carbon-market mechanisms have been widely used by some developed states towards the attainment of their 2008-2012 emissions reduction goals.[108] Furthermore, as discussed below, those mechanisms are expected to be instrumental in some parties' compliance with their KP targets insofar as transfer

---

[105] *Modalities, rules and guidelines for emissions trading under Article 17 of the Kyoto Protocol*, UNFCCC Decision 11/CMP.1 (2005); *Modalities and procedures for a clean development mechanism as defined in Article 12 of the Kyoto Protocol*, UNFCCC Decision 3/CMP.1 (2005); *Guidelines for the implementation of Article 6 of the Kyoto Protocol*, UNFCCC Decision 9/CMP.1 (2005).

[106] 1997 KP, Arts. 6(1)(b), 12 (5)(c).

[107] See *Modalities for the accounting of assigned amounts under Article 7, paragraph 4, of the Kyoto Protocol*, UNFCCC Decision 13/CMP.1 (2005); *Standard electronic format for reporting Kyoto Protocol units*, UNFCCC Decision 14/CMP.1 (2005); *Guidelines for the preparation of the information required under Article 7 of the Kyoto Protocol*, UNFCCC Decision 15/CMP.1 (2005); *Guidelines for national systems under Article 5, paragraph 1, of the Kyoto Protocol*, UNFCCC Decision 19/CMP.1 (2005); *Good practice guidance and adjustments under Article 5, paragraph 2, of the Kyoto Protocol*, UNFCCC Decision 20/CMP.1 (2005); *Issues relating to adjustments under Article 5, paragraph 2, of the Kyoto Protocol*, UNFCCC Decision 21/CMP.1 (2005); *Guidelines for review under Article 8 of the Kyoto Protocol*, UNFCCC Decision 22/CMP.1 (2005); *Terms of service for lead reviewers*, UNFCCC Decision 23/CMP.1 (2005); *Issues relating to the implementation of Article 8 of the Kyoto Protocol – 1*, UNFCCC Decision 24/CMP.1 (2005); and *Issues relating to the implementation of Article 8 of the Kyoto Protocol – 2*, UNFCCC Decision 25/CMP.1 (2005).

[108] *Annual report of the administrator of the international transaction log under the Kyoto Protocol*, note by the secretariat, FCCC/KP/CMP/2011/7 (2011). Annual reports for other years can be viewed at: <http://unfccc.int/kyoto_protocol/registry_systems/itl/items/4065.php> (last visited on 15 August 2012).

and acquisition of emission units during the true-up period following the submission of inventories for the first commitment period is concerned.

2.2.2.2 Compliance

A robust compliance mechanism is key to the success of any Multilateral Environmental Agreement (MEA), and the Kyoto Protocol compliance system is among the most comprehensive and rigorous ones. Its objective is to facilitate, promote and enforce compliance with the commitments under the Protocol.[109] The KP compliance mechanism aims at the restoration of compliance to ensure the environmental integrity of the Protocol.[110] The KP compliance mechanism is unlike any other MEA compliance system because in addition to facilitating and promoting compliance, it plays a robust role in enforcing it.[111] Among the (non-) compliance procedures, it has been described as 'the most similar to a judicial or arbitral one.'[112]

The KP Compliance Committee is an independent international body that provides administrative review of states' implementation of the Protocol. Besides ensuring compliance with the KP emissions reduction targets, the Committee plays a part in securing the accuracy of measurement, reporting, and verification of GHG emissions, and the effective functioning of the carbon-market mechanisms.[113] Whereas it mainly operates through its two branches, namely the Facilitative Branch (FB) and the Enforcement Branch (EB), its members and alternate members form the plenary, and the chair- and vice-chairpersons – the bureau of the Compliance Committee whose official primary function is the allocation of questions of implementation to the appropriate branch.

Fundamental to the KP compliance system is the method of triggering. A question of implementation can be submitted to the Compliance Committee by (1) an expert review team (ERT); (2) a party with respect to itself (self-trigger); or (3) a party with respect to another party (party-to-party trigger).[114]

---

[109] *Procedures and mechanisms relating to compliance under the Kyoto Protocol*, UNFCCC Decision 27/CMP.1 (2005), Ann., Section I (Objective).

[110] *Procedures and mechanisms relating to compliance under the Kyoto Protocol*, UNFCCC Decision 27/CMP.1 (2005), Ann., Section V, para. 6.

[111] R. Lefeber & S. Oberthür, 'Holding countries to account: the Kyoto Protocol's compliance system revisited after four years of experience,' 1 *Climate Law* 133, 2010, p. 134.

[112] T. Treves., The Settlement of Disputes and Non-compliance Procedures, *in* T. Treves *et al.* (Eds.), *Non-compliance Procedures and Mechanisms and the Effectiveness of International Environmental Agreements* (499-518), The Hague: TMC Asser Press, 2009, at 510.

[113] Lefeber & Oberthür 2010, p. 134.

[114] *Procedures and mechanisms relating to compliance under the Kyoto Protocol*, UNFCCC Decision 27/CMP.1, Section VI(1); see also Lefeber & Oberthür 2010, p. 141.

The KP Compliance Committee addresses questions of implementation under the Kyoto Protocol. Procedures of the Compliance Committee relating to the consideration of questions of implementation are subject to detailed prescriptions. The FB is responsible for addressing questions of implementation that fall outside the mandate of the EB. In particular, it is charged with an early-warning function relating to emissions targets prior to the end of the relevant commitment period and methodological and reporting requirements preceding the first commitment period.[115] The function of the FB is to provide advice to promote and facilitate compliance by way of applying a range of soft consequences. To date, the FB has only addressed the 2006 submission by South Africa as Chair of Group 77 and China in respect of several Parties included in Annex I, which was marked by a political stalemate when the Branch was unable to reach an agreement on how to address the question of implementation before it.[116]

The function of the EB is to determine consequences for Parties not meeting certain commitments under the Protocol. The EB is mandated to resolve questions of implementation related to potential cases of non-compliance by industrialized countries with (1) their emissions reduction or limitation targets under Article 3(1) of the Protocol; (2) methodological and reporting requirements under the KP, which include: (a) establishment of national systems for the estimation of anthropogenic emissions by sources and removals by sinks (Art. 5(1)); (b) methodologies for estimating anthropogenic emissions by sources and removals by sinks accepted by the IPCC or agreed by the COP (Art. 5(2)); (c) inclusion in annual inventories of supplementary information for the purposes of ensuring compliance (Art. 7(1)); and (d) guidelines for the preparation of national communications by Annex I parties and modalities for the accounting of assigned amounts adopted by the COP (Arts. 7(4)); and (3) eligibility requirements for participation in the KP market mechanisms under Article 6 (Joint Implementation), 12 (CDM), and 17 (international emissions trading), including reinstatement of eligibility taken away following a prior decision of the EB.[117] The EB is also mandated to resolve any disagreements between the ERT and the relevant party about whether to adjust GHG-emission inventories or concerning a correction to the database for the accounting of assigned amounts.[118]

---

[115] Lefeber & Oberthür 2010, p. 137.

[116] The FB only took decisions not to proceed against Latvia and Slovenia. South Africa's submission is available from: <http://unfccc.int/playground/items/5516.php> (last visited on 17 April 2012); see also Lefeber & Oberthür 2010, pp. 137-138.

[117] *Procedures and mechanisms relating to compliance under the Kyoto Protocol*, UNFCCC Decision 27/CMP.1 (2005), Ann., Section V, para. 4.

[118] *Procedures and mechanisms relating to compliance under the Kyoto Protocol*, UNFCCC Decision 27/CMP.1 (2005), Ann., Section V, para. 5.

Should the EB find a party in non-compliance, it has to apply particular 'consequences' that are associated with each individual type of non-compliance. If a case of non-compliance relates to emissions targets, the EB has to declare the party concerned in non-compliance; deduct 1.3 times the excess tonnes from the party's assigned amount for the second commitment period; request the submission of a compliance action plan; and suspend the party's eligibility to sell emission units. However, the EB will not receive any questions of implementation related to non-compliance with emissions targets before the second half of 2015 since the inventories for the last year of the first commitment period must be submitted by 15 April 2014, and ERTs have up to one year to review them. This is followed by a 100-day true-up period, during which parties to the Protocol may transfer and acquire emission units to achieve compliance with their individual emissions limitation and reduction targets. At present, the EB can only decide on the consequences of non-compliance with the methodological, reporting, and eligibility requirements.[119] The deduction rate applied by the EB amounts to a *de facto* penalty because of the decreasing interest rates since 2001. When during the negotiations of the Compliance Procedures the adjustment rate of 1.3 times the party's excess emissions was agreed upon, the goal was essentially not to penalize but to provide an incentive to comply. It was important, however, to ensure that the defaulting party does not benefit from its non-compliance, which is why the deduction rate had to reflect the opportunity costs of compliance, *i.e.* the possibility of significantly higher returns on the funds earmarked for compliance measures but invested elsewhere would have undermined the KP compliance system.[120] This is why the deduction rate of 1.3 times the party's excess emissions was, at the time of the conclusion of the negotiations on the matter in 2001, a reflection of the compound interest rate on the InterBank market for a five-year period.[121]

If a case of non-compliance concerns methodological and reporting requirements, the EB has to declare the party in question non-compliant and request from it a plan to return to compliance. Should a case of non-compliance relate to eligibility requirements, the EB, in principle, has to suspend the party's eligibility to participate in the carbon-market mechanisms (or, in the case of initial eligibility, decide that a party is not eligible).[122] To date, the EB has addressed questions of implementation in the cases of the following parties: Greece, Canada, Croatia, Bulgaria, Romania, the Ukraine, Lithuania, and Slovakia.[123] All questions of

---

[119] Lefeber & Oberthür 2010, p. 149.
[120] Lefeber & Oberthür 2010, p. 150.
[121] Lefeber & Oberthür 2010, p. 150.
[122] Lefeber & Oberthür 2010, p. 148.
[123] Question of Implementation – Greece, KP Compliance Committee, Final Decision, CC-2007-1-8/Greece/EB (17 April 2008); Question of Implementation – Canada, KP Compliance Committee, Final Decision, CC-2008-1-6/Canada/EB (15 June 2008); Question of Implementation – Croatia, KP Compliance

implementation that have thus far proceeded to the merits have been received from ERTs. The questions of implementation submitted to the EB in relation to those parties, with the exception of Slovakia, only involved methodological and reporting requirements and corresponding eligibility requirements. Therefore, the EB has applied the consequences for those two particular forms of non-compliance (see above). In the case with Slovakia, the EB considered specific functions of Slovakia's national system for the estimation of anthropogenic emissions of GHGs by sources and removals by sinks.[124] It found that Slovakia had in place a national system in accordance with Article 5(1) of the KP. However, '[i]t appear[ed] to the Enforcement Branch that there was a *partial operational impairment* of the performance of some of the specific functions of Slovakia's national system during the review of Slovakia's 2011 annual submission.'[125] The EB concluded that this 'partial operational impairment' resulted in Slovakia's non-compliance with Article 5(1) but not with the eligibility requirements under Articles 6, 12, and 17 of the KP. Because of this conclusion, the EB was able for the first time to apply consequences without suspending the country's eligibility to participate in the KP market mechanisms. Slovakia was declared in non-compliance and was required to develop a plan to return to compliance.[126]

Overall, to date, the EB has been very effective. It has considered questions of implementation in respect of 8 parties, out of which 7 have already returned to compliance, and one party is on its way to achieving compliance.

2.2.3   Developments in the UN Climate Process

Already in 2005, CMP1 held in Montreal, Canada, initiated a process to consider further commitments for industrialized countries beyond 2012 establishing for that

---

Committee, Final Decision, CC-2009-1-8/Croatia/EB (26 November 2009); Question of Implementation – Bulgaria, KP Compliance Committee, Final Decision, CC-2010-1-8/Bulgaria/EB (28 June 2010); Question of Implementation – Romania, KP Compliance Committee, Final Decision, CC-2011-1-8/Romania/EB (27 August 2011); Question of Implementation – Ukraine, KP Compliance Committee, Final Decision, CC-2011-2-9/Ukraine/EB (12 October 2011); Question of Implementation – Lithuania, KP Compliance Committee, Final Decision, CC-2011-3-8/Lithuania/EB (21 December 2011); Question of Implementation – Slovakia, KP Compliance Committee, Preliminary finding, CC-2012-1-7/Slovakia/EB (14 July 2012), confirmed by final decision CC-2012-1-9/Slovakia/EB (17 August 2012).

[124] The EB also considered the disagreement between Slovakia and the ERT on whether to apply adjustments to inventories.

[125] Question of Implementation – Slovakia, KP Compliance Committee, Preliminary finding, CC-2012-1-7/Slovakia/EB (14 July 2012), para. 24, emphasis added.

[126] Question of Implementation – Slovakia, KP Compliance Committee, Preliminary finding, CC-2012-1-7/Slovakia/EB (14 July 2012), confirmed by final decision CC-2012-1-9/Slovakia/EB (17 August 2012).

purpose an *ad hoc* working group on further commitments of Annex I parties.[127] At the same time, in a parallel process, COP11 resolved to engage in a dialogue to analyse long-term cooperative action to address climate change under the Convention through a series of four workshops (the Convention Dialogue).[128]

In 2007, COP13/CMP3 met in Bali, Indonesia. COP13 decided to 'urgently enhance' the implementation of the Convention in order to achieve its ultimate objective.[129] The talks in Bali resulted in the Bali Road Map charting a two-year process moving forward along the two negotiating tracks (AWG-KP – *Ad Hoc* Working Group on Further Commitments for Annex I Parties under the Kyoto Protocol and the AWG-LCA – *Ad Hoc* Working Group on Long-term Cooperative Action under the Convention). The negotiating process launched in Bali aimed at concluding a legally-binding agreement, which was scheduled to complete at COP15 in Copenhagen in 2009. The Bali Action Plan (BAP) adopted under the Convention launched a comprehensive process with a view to reach an agreed outcome and adopt a decision at COP15 focusing on the following five issues: (1) shared vision on long-term cooperative action including a long-term global goal for emissions reduction; (2) enhanced action on mitigation; (3) enhanced action on adaptation including, *inter alia*, consideration of international cooperation; (4) enhanced action on technology development and transfer; and (5) enhanced action on the provision of financial resources.[130] The BAP maintained the differentiation in mitigation commitments between industrialized and developing countries providing for the voluntary nationally appropriate mitigation actions (NAMAs) by developing countries being supported and enabled by technology, financing, and capacity-building by industrialized states.[131]

In 2009, COP15 in Copenhagen did not deliver an agreed outcome mandated by the BAP. During the high-level segment attended by 130 heads of state, a political agreement was negotiated. When the agreement was presented to the plenary, a number of developing states raised concerns about transparency and fairness of the process and refused to adopt it. As a result, COP15 did not adopt the Copenhagen Accord but merely took note of it. Nevertheless, 141 states expressed an intention to be listed as agreeing to the Accord and a number of developed countries provided information on their individual emissions limitation and reduction targets to be met by 2020, and the following year, COP16 took note of those voluntary

---

[127] *Consideration of commitments for subsequent periods for Parties included in Annex I to the Convention under Article 3, paragraph 9*, UNFCCC Decision 1/CMP.1 (2005), paras 1-2.

[128] *Dialogue on long-term cooperative action to address climate change by enhancing implementation of the Convention*, UNFCCC Decision, 1/CP.11 (2005), paras 1, 7(a).

[129] *Bali Action Plan*, UNFCCC Decision 1/CP.13 (2007).

[130] *Bali Action Plan*, UNFCCC Decision 1/CP.13 (2007), para. 1.

[131] *Bali Action Plan*, UNFCCC Decision 1/CP.13 (2007), para. 1(b)(ii).

targets.[132] Many developing countries, too, announced their NAMAs with the aim of achieving 'a deviation in emissions relative to 'business as usual' emissions in 2020.'[133] Notably, the 2020 targets volunteered by industrialized and developing states alike are not legally-binding on either group of states. They may be seen as an expression of the parties' good will; yet, those pledges could not close the mitigation gap between now and 2020 to make up for the difference between the current levels of ambition and what is recommended by science.[134]

On adaptation, the Copenhagen Accord stressed 'the need to establish a comprehensive adaptation programme including international support.' [135] It reiterated the developed countries' intention to 'provide adequate, predictable and sustainable financial resources, technology and capacity-building to support the implementation of adaptation action in developing countries' and stated that 'new and additional, predictable and adequate funding as well as improved access shall be provided to developing countries [...] to enable and support enhanced action' on mitigation, adaptation, technology development and transfer as well as capacity-building.[136]

In 2010, COP16/CMP6 met in Cancun, Mexico, where parties adopted a comprehensive package of decisions under both negotiating tracks. Under the Convention, the Cancun Agreements recognized the need for 'deep cuts in global greenhouse gas emissions' in order to 'hold the increase in global average temperature below 2°C above preindustrial levels' as well as the need to strengthen the long-term global goal 'on the basis of the best available scientific knowledge, including in relation to a global average temperature rise of 1.5°C.' [137] As mentioned earlier, the Cancun Agreements also took note of NAMAs communicated by industrialized as well as developing countries following the Copenhagen Accord.

---

[132] *Cancun Agreements: Outcome of the work of the* Ad Hoc *Working Group on Long-term Cooperative Action under the Convention,* UNFCCC Decision 1/CP.16, para. 36; for specific targets per country see *Compilation of economy-wide emission reduction targets to be implemented by Parties included in Annex I to the Convention,* revised note by the secretariat, FCCC/SB/2011/INF.1/Rev.1 (2011).

[133] *Cancun Agreements: Outcome of the work of the* Ad Hoc *Working Group on Long-term Cooperative Action under the Convention,* UNFCCC Decision 1/CP.16, para. 48; for a compilation of NAMAs per developing country, see *Compilation of information on nationally appropriate mitigation actions to be implemented by Parties not included in Annex I to the Convention,* note by the secretariat, FCCC/AWGLCA/2011/INF.1 (2011).

[134] *UNEP Emissions Gap Report,* Technical Summary (2010), p.10.

[135] *Copenhagen Accord,* UNFCCC Decision 2/CP.15 (2009), para. 1.

[136] *Copenhagen Accord,* UNFCCC Decision 2/CP.15 (2009), paras 3, 8.

[137] *Cancun Agreements: Outcome of the work of the* Ad Hoc *Working Group on Long-term Cooperative Action under the Convention,* UNFCCC Decision 1/CP.16 (2010), para. 4.

With regard to adaptation, COP16 adopted a decision on enhanced action and international cooperation on adaptation. For that purpose, the Cancun Adaptation Framework was established as well as a process to enable least developed countries (LDCs) 'to formulate and implement national adaptation plans' (NAPs).[138] Also, an Adaptation Committee was set up in order 'to promote the implementation of enhanced action on adaptation [...] under the Convention.'[139] Further, the COP 'requested' developed countries to provide developing countries with 'long-term, scaled-up, predictable, new and additional finance, technology and capacity-building [...] to implement urgent, short-, medium- and long-term adaptation actions, plans, programmes and projects at the local, national, subregional and regional levels, in and across different economic and social sectors and ecosystems.'[140] At the same time, developed countries undertook a collective commitment of providing new and additional resources 'approaching USD 30 billion for the period 2010-2012, with a balanced allocation between adaptation and mitigation' and with funding for adaptation being prioritized for the most vulnerable developing countries, such as LDCs, small island developing states (SIDS), and Africa.[141] Since these short-term resources are intended for financing both adaptation and mitigation, the actual figure for adaptation efforts is considerably lower.

Under the Protocol, parties agreed to aim to complete negotiations on a second commitment period as soon as possible and 'in time to ensure that there is no gap between the first and second commitment periods.'[142] The CMP urged industrialized countries to raise the level of ambition with a view to reducing their GHG emissions in accordance with the range indicated in the IPCC Fourth Assessment Report.[143] But it was not until COP17/CMP7 that agreement was reached on another commitment period and not until COP18/CMP8 that the KP was amended (see below).

---

[138] *Cancun Agreements: Outcome of the work of the* Ad Hoc *Working Group on Long-term Cooperative Action under the Convention*, UNFCCC Decision 1/CP.16 (2010), paras 13-15.

[139] *Cancun Agreements: Outcome of the work of the* Ad Hoc *Working Group on Long-term Cooperative Action under the Convention*, UNFCCC Decision 1/CP.16 (2010), para. 20.

[140] *Cancun Agreements: Outcome of the work of the* Ad Hoc *Working Group on Long-term Cooperative Action under the Convention*, UNFCCC Decision 1/CP.16 (2010), para. 18.

[141] *Cancun Agreements: Outcome of the work of the* Ad Hoc *Working Group on Long-term Cooperative Action under the Convention*, UNFCCC Decision 1/CP.16 (2010), para. 95.

[142] *Cancun Agreements: Outcome of the work of the* Ad Hoc *Working Group on Further Commitments for Annex I Parties under the Kyoto Protocol at its fifteenth session*, UNFCCC Decision 1/CMP.6 (2010), para. 1.

[143] *Cancun Agreements: Outcome of the work of the* Ad Hoc *Working Group on Further Commitments for Annex I Parties under the Kyoto Protocol at its fifteenth session*, UNFCCC Decision 1/CMP.6 (2010), para. 4.

COP17/CMP7 met in Durban, South Africa, in 2011. Under the Convention track, the Durban negotiations resulted in the establishment of an *Ad Hoc* Working Group on the Durban Platform for Enhanced Action (ADP) with a mandate to develop 'a protocol, another legal instrument or an agreed outcome with legal force under the Convention applicable to all Parties' no later than 2015, with the new instrument coming into effect by 2020.[144] Although this formulation appears to allow a lot of flexibility as to the mandated outcome of the ADP's work, in practice, options are limited to a protocol, a new international agreement or a set of COP decisions. COP17 decided that, informed by the IPCC Fifth Assessment Report to be released in 2013-2015, the process shall raise the level of ambition. Parties also launched a work plan to enhance the mitigation ambition and explore options for closing the ambition gap between now and 2020.[145] The language of the decision establishing the Durban Platform is broad enough to account for a number of possibilities but appears to suggest that the new agreement will include industrialized and developing states alike. The exact mitigation commitments are yet to be formulated but it is significant that the decision on the Durban Platform contains no reference to the principles of equity and common but differentiated responsibilities and respective capabilities. The first session of the ADP was held in Bonn in May 2012, during which the officers were elected and the agenda was adopted.

With regard to adaptation, COP17 further advanced the implementation of the Cancun Adaptation Framework. Parties agreed on modalities, procedures, and composition of the Adaptation Committee and modalities and guidelines for NAPs. Notably, the Conference adopted a decision on approaches to address loss and damage associated with climate change impacts in developing countries that are particularly vulnerable to the adverse effects of climate change and agreed on activities to be undertaken under the work programme on loss and damage.[146] In addition to the existing funding options for climate change adaptation (the Global Environmental Facility Trust Fund, Least Developed Countries Fund, Special Climate Change Fund, and Adaptation Fund), the COP in Durban also designated the Green Climate Fund (GCF) as an operating entity of the financial mechanism under the Convention. While the GCF could 'become the main global fund for climate finance,'[147] contributions to the fund by industrialized states are voluntary

---

[144] *Establishment of an* Ad Hoc *Working Group on the Durban Platform for Enhanced Action*, UNFCCC Decision 1/CP.17 (2011), paras 1-2, 5.

[145] *Establishment of an* Ad Hoc *Working Group on the Durban Platform for Enhanced Action*, Decision 1/CP.17 (2011), paras 6-7.

[146] *Work programme on loss and damage*, UNFCCC Decision 7/CP.17 (2011).

[147] S. Aguilar *et al.*, Summary of the Durban Climate Change Conference, *Earth Negotiations Bulletin*, 28 November – 9 December 2011, p. 4, quoting the COP17 President Nkoana-Mashabane.

and the funding is earmarked for both adaptation and mitigation, and includes capacity-building and technology development and transfer.[148]

During simultaneous negotiations under the Protocol, parties were able to reach an agreement on a period; however the new commitment period was not adopted by amendment to the Protocol. The second commitment period, it was agreed, shall begin on 1 January 2013 and end on 31 December 2017 or 31 December 2020.[149]

In 2012, COP18/CMP8 met in Doha, Qatar. Parties adopted a package of decisions known as the Doha Climate Gateway. Under the Convention, the AWG-LCA completed its work, as scheduled. Following an informal meeting in Bangkok, Thailand, in August – September 2012, the first ADP session was resumed. Parties continued discussions under the two workstreams identified in Bonn, namely the post-2020 regime and ways to bridge the pre-2020 mitigation ambition gap.

In relation to adaptation, parties decided to develop at COP19 institutional arrangements, such as an international mechanism, to address loss and damage in developing countries that are particularly vulnerable to the adverse effects of climate change.[150]

Under the Protocol, parties adopted an amendment to the KP, in accordance with which the second commitment period beginning on 1 January 2013 will end on 31 December 2020.[151] A new Article 3, paragraph 1*bis*, provides for the objective of reducing, individually or jointly, aggregate emissions by industrialized countries by at least 18 per cent below 1990 levels in the commitment period 2013-2020. The amendment contains a new Annex B setting out developed countries' individual quantified emissions limitation and reduction commitments for the second commitment period. It was agreed that each party with a quantified emissions limitation and reduction commitment will revisit its commitment by 2014 at the latest and may increase its ambition in line with an aggregate reduction of GHG emissions of at least 25-40 per cent below 1990 levels by 2020.[152]

---

[148] *Launching the Green Climate Fund*, UNFCCC Decision 3/CP.17 (2011), para. 25; Ann., paras 37-38.

[149] *Outcome of the work of the* Ad Hoc *Working Group on Further Commitments for Annex I Parties under the Kyoto Protocol at its seventh session*, UNFCCC Decision 1/CMP.7 (2011), para. 1.

[150] *Approaches to address loss and damage associated with climate change impacts in developing countries that are particularly vulnerable to the adverse effects of climate change to enhance adaptive capacity*, FCCC/CP/2012/L.4/Rev.1 (UNFCCC decision number not available at the time of writing), 8 December 2012, para. 9.

[151] *Outcome of the work of the* Ad Hoc *Working Group on Further Commitments for Annex I Parties under the Kyoto Protocol*, FCCC/KP/CMP/2012/L.9 (UNFCCC decision number not available at the time of writing), 8 December 2012, para. 4.

[152] *Outcome of the work of the* Ad Hoc *Working Group on Further Commitments for Annex I Parties under the Kyoto Protocol*, FCCC/KP/CMP/2012/L.9 (UNFCCC decision number not available at the time of writing), 8 December 2012, para. 7.

It was agreed that the market-based mechanisms under the Protocol continue in the second commitment period. The CMP clarified that developed and developing countries continue to be able to participate in CDM projects; however, only developed countries with a quantified emissions limitation and reduction commitment can transfer and acquire CERs. With respect to JI and emissions trading, the CMP decided that only industrialized countries with a quantified emissions limitation and reduction commitment shall be eligible to transfer and acquire CERs, AAUs, ERUs, and RMUs valid for the second commitment period.

2.3 Concluding Remarks

In Durban, all industrialized countries and almost fifty developing countries reaffirmed their previously announced mitigation pledges for 2020. Together those pledges count for 80 per cent of the global emissions. However, despite the perceived progress on further emissions reduction and limitation commitments, the 2020 targets are not binding for developing countries, and two developed countries and one EIT – Canada, Japan, and the Russian Federation – have indicated that they have no intention of participating in the KP second commitment period.[153] New Zealand has indicated that, while remaining party to the KP, it will be taking a quantified economy-wide emissions reduction target in the period 2013 to 2020 under the Convention. The largest emitter among industrialized nations – the US – has never been bound by the Protocol's provisions as it never became party to it. The KP, as amended in Doha, contains no legally binding targets for developing countries, which means that major emitters such as China (currently the world's largest emitter), India, Brazil, and South Africa are under no legal obligation on the basis of the Protocol to take mitigation measures towards GHG emissions reduction or limitation. Following COP18/CMP8, only about 15 per cent of the global emissions are currently covered by the Kyoto Protocol. The negotiations towards a new all-inclusive agreement under the Durban Platform will not produce an outcome before 2015 and whatever document is agreed upon will not take effect until 2020. Even so, it is far from certain that developing countries will agree to binding targets and without the participation of developing states, it is unlikely that the US will undertake legally binding emissions reduction commitments. Yet, according to most scenarios, the global emissions would have to peak by 2015 and then decline for the average global temperatures to remain within the confines of a 2°C rise.[154]

---

[153] *Outcome of the work of the* Ad Hoc *Working Group on Further Commitments for Annex I Parties under the Kyoto Protocol at its seventh session*, UNFCCC Decision 1/CMP.7 (2011), Ann. 1.

[154] IPCC *AR4, Synthesis Report* (2007), p. 67.

Seeing that the international climate regime does not obligate developing states to take mitigation measures and fewer industrialized countries have undertaken emissions limitation and reduction commitments under the KP second commitment period than in the 2008-2012 commitment period (Canada has withdrawn from the KP altogether), the ultimate objective of the Convention is becoming more difficult to attain. While the progress of climate negotiations is slow, international mitigation needs to become considerably more ambitious to avoid dangerous climate destabilization. According to the UNEP's Emissions Gap Report, the expected gap between emission levels consistent with a 2°C limit and those resulting from the non-binding pledges made following the Copenhagen Accord would range from 5 to 9 GtCO$_2$eq (gigatonnes of carbon dioxide equivalent), depending on how the pledges are implemented.[155] Therefore, it is imperative to ensure that all states take mitigation measures. In this context, interstate liability – or, more precisely, state responsibility – can play an important role (see Chapter 5). The next chapter examines the different modalities of interstate liability in international law and identifies a legal framework for compelling states to take mitigation measures.

---

[155] *UNEP Emissions Gap Report*, Technical Summary (2010), p. 10.

3 Identification of International Liability Regimes Relevant to Climate Change-Related Damage

3.1 Introduction

Interstate liability can be established under primary and secondary norms of international law. In the law of state responsibility, international obligations regulating the conduct of international actors in a specific sector of interstate relations are referred to as primary rules of international law. The determination of legal consequences of a state's failure to fulfil those obligations is often not governed by the primary rules of a particular sector, but rather by the secondary rules of international law. When arising from primary norms of international law, interstate liability may be more accurately described by the generally accepted term 'state liability' and is triggered when lawful acts of a state lead to harm in another state's territory. In contrast, state responsibility is engaged under secondary rules of international law codified by the International Law Commission (ILC) in its Articles on Responsibility of States for Internationally Wrongful Acts (Articles on State Responsibility or ILC Articles). It is distinct from state liability in that it is predicated on the existence of an internationally wrongful act.

The present chapter is based first and foremost on the differentiation between state liability on the one hand and state responsibility on the other.[156] Often understated in academic literature, this distinction is fundamental to any discussion of liability at the interstate level. While the former is grounded in specific primary rules of international law and is applied in situations when lawful conduct of a state has resulted in harm to another state, the latter pertains to secondary norms regulating the consequences of state acts that are considered internationally wrongful.

In dealing with state liability, the chapter addresses various liability mechanisms under MEAs and distinguishes four conceptual approaches based on the nature of international obligations the relevant primary norms create. Depending on the approach adopted by a particular legal regime, state liability may give rise to: (1) the obligation to pay compensation; (2) the obligation to negotiate a redress settlement; (3) the obligation to ensure prompt, adequate, and effective compensation; or (4) the obligation to take response action. After analysing the various approaches to state liability under primary rules of international law, the

---

[156] See *e.g.* M. Fitzmaurice, International Responsibility and Liability, *in* D. Bodansky, J. Brunnée & E. Hey (Eds.), *The Oxford Handbook of International Environmental Law* (1010-1035). New York: Oxford University Press, 2007.

chapter shifts its focus to secondary norms of international law that regulate matters of state responsibility following commission of an internationally wrongful act.

3.2 State Responsibility vs. State Liability

Some international legal regimes are strengthened by liability provisions designed to remedy instances when lawful conduct results in damage. Examples include: the 1972 Convention on International Liability for Damage Caused by Space Objects (Space Liability Convention); the 1989 Basel Convention on the Control of Transboundary Movements of Hazardous Wastes and their Disposal;[157] and the 2010 Nagoya-Kuala Lumpur Supplementary Protocol on Liability and Redress to the CBD. Those provisions are based on different approaches to state liability and, as will soon be demonstrated, are often limited in scope. All state liability mechanisms, regardless of the approach they are based on, share an important characteristic: they all address the injurious consequences of acts that are lawful. As mentioned earlier, the determination of the legal consequences of a state's conduct that is internationally wrongful is governed by the law of state responsibility. State responsibility requires a breach of an international obligation while state liability arises out of 'harm alone' whereby proof of an internationally wrongful act is not a requirement.[158] Also, Kiss & Shelton note that nowadays, 'international environmental law has come to distinguish [1] responsibility, which arises upon breach of an international obligation, and [2] liability for the injurious consequences of lawful activities.'[159] Lefeber makes the same distinction by describing state responsibility as 'liability *ex delicto*' and referring to state liability as 'liability *sine delicto*.'[160] In other words, state responsibility originates from acts that are internationally wrongful and as such may be distinguished from 'liability for the deleterious effects of lawful acts.'[161]

---

[157] See 1999 Basel Protocol on Liability and Compensation for Damage Resulting from the Transboundary Movement of Hazardous Wastes and their Disposal, Doc. UNEP/CHW.1/WG/1/9/2 (1999).

[158] C. Hoss, State Responsibility, Liability and Environmental Protection, *in* R. Wolfrum, C. Langenfeld & P. Minnerop (Eds.), *Environmental Liability in International Law: Towards a Coherent Conception* (455-494). Berlin: Erich Schmidt Verlag, 2005, at 455.

[159] A. Kiss & D. Shelton, Strict Liability in International Environmental Law, *in* T.M. Nndiaye & R. Wolfrum (Eds.), *Law of the Sea, Environmental Law and Settlement of Disputes. Liber Amicorum Judge Thomas A. Mensah* (1131-1151); Leiden: Martinus Nijhoff Publishers, 2007, at 1132.

[160] R. Lefeber, *Transboundary Environmental Interference and the Origin of State Liability*, The Hague: Kluwer Law International, 1996, p. 147.

[161] A. Kiss & D. Shelton, *Guide to International Environmental Law*, Leiden: Martinus Nijhoff Publishers, 2007, p. 19.

In international law, states' obligations under treaties or international custom are referred to as primary rules, whereas the consequences of a state's failure to act in accordance with its international obligations are determined by secondary rules of international law, or the law of state responsibility. The law of state responsibility has three main functions: (1) preventive function, aimed at dissuasion from engaging in conduct that could result in an internationally wrongful act; (2) enforcement of primary international obligations, or corrective function; and (3) compensatory, or reparative, function aimed at a restoration of the *status quo ante* or compensation for damages caused by an internationally wrongful act. This 'triple function' of the international law of state responsibility is reflected in the Articles on Responsibility of States for Internationally Wrongful Acts drafted by the ILC.[162]

The ILC's work on the Articles dates back to 1963 when the Commission decided to limit its work to secondary rules as codification of secondary as well as primary obligations appeared to be an impossible task due to the 'infinitely varying nature' of the latter.[163] The distinction between primary and secondary rules is fundamental to the Articles. It was put forward by Special Rapporteur Ago who considered it impossible to move forward without distinguishing between those two sets of rules. Ago felt that primary rules and the obligations they imposed were inherently different from secondary rules that determined whether or not the obligation set by a particular primary rule had been violated. To use Crawford's explanation, 'the key idea is that a breach of a primary obligation gives rise, immediately by operation of the law of state responsibility, to a secondary obligation or series of such obligations' and the Articles 'specify the default rules that determine when a breach occurs and, in general, the content of the resulting secondary obligations.'[164] The breach of a primary obligation, in other words, 'gives rise to a new legal regime, that of state responsibility, that contains its own distinctive set of duties and rights.'[165] Today, the term *state responsibility* is 'widely used to denote secondary rules, following the decision of the ILC to limit its articles on state responsibility to these.'[166]

Even though the focus on secondary rules of international law was intended to facilitate the ILC's work, it took the Commission more than forty years to complete

---

[162] Lefeber 1996, p. 313; see also Hoss 2005, p. 455.

[163] See 2001 Articles on Responsibility of States for Internationally Wrongful Acts, ILC Report on the work of its 53rd session, A/56/10, YILC, vol. II, Part Two; Fitzmaurice 2007, p. 1016.

[164] D. Bodansky, J.R. Crook & J. Crawford, 'The ILC's articles on responsibility of states for internationally wrongful acts: a retrospect,' 96 *AMJIL* 874, 2002, at 876.

[165] M.A. Drumbl, *Trail Smelter* and the International Law Commission's Work on State Responsibility for Internationally Wrongful Acts and State Liability, *in* R.M. Bratspies & R.A. Miller (Eds.), *Transboundary Harm in International Law* (85-98), New York: Cambridge University Press, 2006, at 86.

[166] Fitzmaurice 2007, p. 1011.

the draft. It was finalized in August 2001 and the UN General Assembly took note of it in its Resolution 56/83 in December 2001.[167] The Articles on Responsibility of States for Internationally Wrongful Acts focus on 'the general conditions under international law for the State to be considered responsible for wrongful actions or omissions, and the legal consequences which flow therefrom;'[168] *i.e.* they apply to all areas of international law. While dealing only with conduct that is internationally wrongful, the Articles cover 'the whole field of the international obligations of States, whether the obligation is owed to one or several States, to an individual or group, or to the international community as a whole.'[169] It is significant that in addressing state responsibility *vis-à-vis* one state, a group of states or the entire international community, the Articles could provide 'an important mechanism to cultivate responsibility for state conduct that breaches obligations to protect common concerns of humanity – such as biodiversity and the atmosphere.'[170] In spite of the fact that state responsibility does not cover the liability of private actors who are largely responsible for pollution as such, it has played an increasingly important role in environmental law.[171] It is submitted that the Articles' framework could also be extended to encompass interstate responsibility for the injurious consequences of climate change (see Chapter 5).

3.2.1 State Liability

If, in its general aspects, international law is relatively straightforward on the issue of state responsibility arising out of acts that are considered internationally wrongful, it is far less uniform in cases when harm is caused by a lawful act of the state. Depending on the nature of the regulated activity, situations when harm is caused by the state's lawful acts are governed by a number of state liability regimes. The current section deals with the various conceptual approaches to state liability in international law and evaluates the suitability of those approaches for addressing climate change-related damage based on the following criteria, as appropriate: legal

---

[167] *Responsibility of States for internationally wrongful acts*, GA Res. 56/83 (28 January 2002); see also Kiss & Shelton 2007b, p. 1135.

[168] 2001 Articles on Responsibility of States for Internationally Wrongful Acts, ILC Report on the work of its 53rd session, A/56/10, YILC, vol. II, Part Two, general commentary, p. 59, para. 1.

[169] 2001 Articles on Responsibility of States for Internationally Wrongful Acts. ILC Report on the work of its 53rd session. A/56/10, YILC, vol. II, Part Two, general commentary, p. 62, para. 5.

[170] Drumbl, 2006, p. 89.

[171] *E.g. Legality of the Threat or Use of Nuclear Weapons,* Advisory Opinion of 8 July 1996, 1996 ICJ Rep. 226; *Case Concerning the Gabčíkovo-Nagymaos Project (Hungary v. Slovakia),* Judgment of 25 September 1997, 1997 ICJ Rep. 7; *Case Concerning Pulp Mills on the River Uruguay (Argentina v. Uruguay),* Judgment of 20 April 2010, 2010 ICJ Rep. 14; *Responsibilities and Obligations of States with Respect to Activities in the Area,* Advisory Opinion of 1 February 2011, 2011 ITLOS Rep. 10.

feasibility, political feasibility, conceptual fittingness, and systemic compatibility. As mentioned in the previous section, state liability is distinct from state responsibility in that it does not depend on wrongfulness and is triggered when harm is caused by activities permitted by a state that are lawful.

The approach to state liability addressed first involves the obligation to pay compensation. It dates back to the 1960s when the proliferation of space and nuclear activities alerted mankind to the new risks associated with the administration of those ultra-hazardous activities. Although liability for the operation of nuclear power plants and transportation of nuclear substances is mainly subject to civil liability regimes, [172] some international agreements governing ultra-hazardous activities have opted for state liability. [173] The section below provides an analysis of the obligation to pay compensation based on the example of the 1972 Convention on International Liability for Damage Caused by Space Objects. The second approach refers to the obligation to negotiate a redress settlement and is reflected, for instance, in the 1997 Convention on the Law of Non-navigational Uses of International Watercourses, considered in Section 3.2.1.2. The third approach is concerned with the obligation to ensure prompt, adequate, and effective compensation. This approach is chiefly adopted by international civil liability regimes; however limited possibilities for state liability are also envisaged by the ILC 2001 Principles on Allocation of Loss in the Case of Transboundary Harm Arising out of Hazardous Activities, which are analysed in Section 3.2.1.3. The fourth approach imposes on states the obligation to require the appropriate operators to take response action in the event of damage. Examples of international legal instruments relying on this approach include the 2005 Annex VI to the 1991 Protocol on Environmental Protection to the Antarctic Treaty (Liability Arising from Environmental Emergencies) and the 2010 Nagoya-Kuala Lumpur Supplementary Protocol on Liability and Redress to the CBD, an examination of which will conclude the section on state liability.

---

[172] *E.g.* 1960 OECD Convention on Third Party Liability in the Field of Nuclear Energy, UNTS, vol. 956, p. 263 (1974); 1963 IAEA Convention on Civil Liability for Nuclear Damage, UNTS, vol. 1063, p. 265 (1977); on bilateral agreements on civil liability in the nuclear field, see Lefeber 1996, pp. 240-241.

[173] *E.g.* 1972 Convention on International Liability for Damage Caused by Space Objects United Nations, Treaties and Principles on Outer Space, Text of treaties and principles governing the activities of States in the exploration and use of outer space, adopted by the United Nations General Assembly, ST/SPACE/11, United Nations, New York, 2002, p. 13; 1987 CMEA Convention on Liability for Damage Caused by Radiological Accidents in International Carriage of Irradiated Nuclear Fuel from Nuclear Power Plants, unpublished, for references see International Liability for Injurious Consequences Arising out of Acts not Prohibited by International Law (Prevention of Transboundary Damage from Hazardous Activities), Doc. A/CN.4/501, YILC, vol. II, Part One (1999), pp. 123-124; see also Lefeber 1996.

3.2.1.1 Obligation to Pay Compensation

In environmental law, strict liability of states is usually associated with ultra-hazardous activities involving significant risk – activities that are considered 'especially new or dangerous.'[174] The concept of strict liability of states has been developed to facilitate the recovery of compensation for harm caused by ultra-hazardous activities of nuclear and space-exploration nature.[175] The 1967 Treaty on Principles Governing the Activities of States in the Exploration and Use of Outer Space (Outer Space Treaty) provides that

> [e]ach State Party to the Treaty that launches or procures the launching of an object into outer space, including the Moon and other celestial bodies, and each State Party from whose territory or facility an object is launched, is internationally liable for damage to another State Party to the Treaty or to its natural or juridical persons by such object or its component parts on the Earth, in air space or in outer space, including the moon and other celestial bodies.[176]

The 1972 Space Liability Convention expands on the liability rules of the Outer-Space Treaty and provides a legal framework for the settlement of claims associated with harm arising out of space activities. Under the Space Liability Convention, the compensatory remedy is available to the injured state under an absolute liability scheme in cases of damage caused by space objects on the surface of the Earth or to aircraft in flight (Art. II). In cases of damage to a space object of one launching state by a space object of another launching state elsewhere than on the surface of the Earth, liability of the latter state is fault-based (Art. III). Specifically, the liability of the launching state extends to 'its fault or the fault of persons for whom it is responsible.' Two launching states (also in cases of joint launching) are jointly and severally liable for damage caused to a third state (Arts. IV(1) and V(1)). Their liability is absolute if the damage has been caused to a third state on the surface of the Earth or to aircraft in flight; in cases of damage occurring elsewhere, liability to third states is based on fault (Arts. IV(1)(a-b)).

---

[174] P. Birnie, A. Boyle & C. Redgwell, *International Law & the Environment*, New York: Oxford University Press, 2009, p. 218; Kiss & Shelton 2007a, p. 24; Kiss & Shelton 2007b, p. 1135; Lefeber 1996, p. 150.

[175] Lefeber 1996, p. 159.

[176] 1967 Treaty on Principles Governing the Activities of States in the Exploration and Use of Outer Space, including the Moon and Other Celestial Bodies, United Nations Treaties and Principles on Outer Space, Text of treaties and principles governing the activities of States in the exploration and use of outer space, adopted by the United Nations General Assembly, ST/SPACE/11, United Nations, New York, 2002, p. 3, Art. VII.

As far as the amount of compensation payable is concerned, the Space Liability Convention follows the principle of full compensation.[177] Article XII provides:

> The compensation which the launching State shall be liable to pay for damage under this Convention shall be determined in accordance with international law and the principles of justice and equity, in order to provide such reparation in respect of the damage as will restore the person, natural or juridical, State or international organization on whose behalf the claim is presented to the condition which would have existed if the damage had not occurred.

If a claim is not settled through diplomatic negotiations in accordance with Article IX, a Claims Commission may be established (Art. XIV). The Space Liability Convention does not set any maximum liability limits and the Claims Commission does not only decide on the merits of the claim but also determines the amount of compensation payable (Art. XVIII). It must be stressed, however, that the Space Liability Convention supports state liability only *vis-à-vis* other states and does not grant procedural access to victims.

The Space Liability Convention is subject to a number of limitations as its scope only extends to damage caused by space objects. Such occurrences are rather uncommon and no claims have been brought under the Convention to date. However, one incident that almost resulted in a formal compensation claim must be mentioned. In January 1978, Cosmos 954, a Soviet satellite carrying a nuclear reactor, re-entered the Earth's atmosphere, partially disintegrated, and left a trail of radioactive debris over a remote territory in northern Canada.[178] Canada undertook extensive clean-up activities and, in January 1979, presented to the Soviet Union a formal claim for compensation asserting that the USSR was absolutely liable to pay compensation under Article II of the Space Liability Convention. Canada demanded six million dollars in compensation. The Soviet Union rejected the claim and, after lengthy negotiations between the two governments, in 1981 a protocol was concluded in accordance with which the USSR agreed to pay compensation of three million dollars 'in full and final settlement of all matters connected with the disintegration of the Soviet satellite Cosmos 954 in January 1978.'[179] Although the ultimate resolution of the Cosmos 954 case was not based on the Convention as it

---

[177] C. Langenfeld & P. Minnerop, Environmental Liability Provisions in International Law, *in* R. Wolfrum, C. Langenfeld & P. Minnerop (Eds.), *Environmental Liability in International Law: Towards a Coherent Conception* (3-158), Berlin: Erich Schmidt Verlag, 2005, at 71.

[178] On the Cosmos 954 incident, see K. Böckstiegel, Case Law on Space Activities, *in* N. Jasentuliyana (Ed.), *Space Law: Development and Scope* (206-218), Westport: Praeger Publishers, 1992, at 206; Langenfeld & Minnerop 2005, pp. 70-71.

[179] Böckstiegel 1992, p. 206.

was settled through the payment of a lump sum, the incident revealed a regulatory gap with regard to the use of nuclear power sources in outer space. The applicability of the 1972 Space Liability Convention to nuclear power sources, as well as the inclusion of 'expenses for search, recovery and clean-up operations' in cases of damage caused by space objects, were subsequently clarified in the Principles Relevant to the Use of Nuclear Power Sources in Outer Space adopted by the UN General Assembly in 1992.[180]

Next to the Space Liability Convention, the only other treaty that imposes absolute liability on states is the obsolete 1987 Council for Mutual Economic Assistance (CMEA) Convention on Liability for Damage Caused by Radiological Accidents in International Carriage of Irradiated Nuclear Fuel from Nuclear Power Plants.[181] Other treaties concerned with nuclear accidents as well as treaties governing oil pollution at sea do not follow this approach but rather opt for civil liability regimes.[182]

In international law, treaties envisaging unlimited compensation in an absolute liability context are extremely rare, and the Space Liability Convention is the only example of an operational international treaty creating an obligation to pay compensation as a primary rule of state liability in its pure form. Under the Convention, the absolute obligation to pay compensation is not based on wrongfulness but is triggered by the occurrence of damage leading to liability *sine delicto*. In the 1970s, when states had a monopoly over operation and launching of space objects, such an approach seemed sensible. With the space activities' continuing spill into the private sector, it would be unlikely for the Space Liability Convention to take the same shape today as states appear 'willing to accept liability for their own conduct, but not for that of private actors.'[183]

---

[180] *Principles Relevant to the Use of Nuclear Power Sources in Outer Space*, UN Res. 47/68 (1992), Principles 9(1) and 9(3).

[181] 1987 CMEA Convention on Liability for Damage Caused by Radiological Accidents in International Carriage of Irradiated Nuclear Fuel from Nuclear Power Plants, unpublished, see footnote 168 *supra*.

[182] *E.g.* 1960 OECD Convention on Third Party Liability in the Field of Nuclear Energy; 1962 Convention on the Liability of Operators of Nuclear Ships, 57 AJIL 268 (1963); 1963 IAEA Convention on Civil Liability for Nuclear Damage, UNTS, vol. 1063, p. 265 (1977); 1969 IMCO International Convention on Civil Liability for Oil Pollution, UNTS, vol. 973, p. 3 (1975); 1971 Convention Relating to Civil Liability in the Field of Maritime Carriage of Nuclear Material, UNTS, vol. 974, p. 255 (1975); see also Birnie, Boyle & Redgwell 2009, p. 218.

[183] Kiss & Shelton 2007a, p. 29; Kiss & Shelton 2007b, p. 1140.

3.2.1.1.1 Obligation to Pay Compensation and Climate Change-Related Damage

In principle, the approach to liability based on the obligation to pay compensation would be an attractive option as far as potential claimant states are concerned. Conceptually, it is not inconceivable to seek compensation for damage caused through the emission of GHGs. However, most of GHG emissions contributing to global warming originate from the private sector, and, politically, states are reluctant to accept absolute liability for the conduct of private actors.[184] The private actor issue also makes absolute liability of states systemically incompatible with climate change-related damage.

Next, from a legal point of view, the absolute liability approach is unsuitable for addressing climate change-related damage due to causation challenges. The difficulty of tracing climate change-related damage in one country to GHG emissions originating in another, coupled with the fact that no country is carbon-neutral, would prevent the imposition of absolute liability on any one state because every state must be deemed to have contributed to climate change. Also, in environmental law, the absolute liability standard is usually associated with activities carrying significant risk. In and of itself, the emission of GHGs is not risky, nor is it especially new or particularly dangerous. GHGs are only capable of causing the warming of the Earth's atmosphere in great concentrations that are achieved over a considerable period of time through emissions from a great many sources.

3.2.1.2 Obligation to Negotiate a Redress Settlement

The obligation to negotiate a redress settlement has been considered by the ILC in its work on 'international liability for injurious consequences arising out of acts not prohibited by international law.' The obligation to negotiate in good faith on the reparation of transboundary harm in the absence of an internationally wrongful act was first introduced by the ILC's Special Rapporteur Robert Quentin Quentin-Baxter who saw it as a substitute for the duty to prevent. In his view, liability would only arise when the harm has occurred and the causal link can be demonstrated. According to Quentin-Baxter, negotiations should be guided by 'shared expectations' of the source state and the affected state, and failure to reach a settlement would necessitate referral of the dispute to a dispute settlement procedure. Later on, his successor Julio Barboza similarly proposed that liability

---

[184] For a discussion of state practice in respect of the no-act-of-the-state exemption, see Lefeber 1996, pp. 159-187, especially p. 181.

disputes be resolved through the procedural obligation to negotiate to determine the legal consequences of harm bearing in mind that the harm must, in principle, be fully compensated. If transboundary harm occurs, diligent behaviour does not exempt the source state from the duty to negotiate with affected states. Ultimately, Barboza, as well as his successor Rao, embraced the obligation to ensure prompt, adequate, and effective compensation as a consequence of transboundary harm in the absence of an internationally wrongful act (see Section 3.2.1.3).

Thus, there are few sources of international law providing support for the obligation to negotiate a redress settlement. The 1997 Convention on the Law of the Non-navigational Uses of International Watercourses, adopted when the ILC was still pursuing the obligation to negotiate a redress settlement, is one of them.[185] Under the Convention, the obligation to negotiate a redress settlement is aimed at the reparation of transboundary harm when such harm has occurred despite the source state's diligent conduct. The due diligence obligation to prevent is reflected in Article 7(1) whereby in utilizing an international watercourse in their territories, watercourse states shall 'take all appropriate measures to prevent the causing of significant harm' to other watercourse states. Should significant harm be caused in spite of diligent behaviour on the part of the source state, the latter shall, in consultation with the affected state, take all appropriate measures to eliminate and mitigate such harm and, where appropriate, discuss the question of compensation (Art. 7(2)). However, although states may be duty-bound to consult, they can neither be obligated to reach an agreement, including on compensation, nor can they be required to refer the dispute to a third party for settlement. In the event of a dispute concerning the interpretation or application of the Convention, the dispute settlement procedure is triggered, in accordance with which, if unable to reach agreement by negotiation requested by one of them, parties to the dispute may jointly have recourse to good offices, mediation and conciliation, or even arbitration or adjudication (Art. 33(2)). All those means of dispute settlement are subject to agreement by the state and therefore cannot be considered compulsory.

3.2.1.2.1   Obligation to Negotiate a Redress Settlement and Climate Change-Related Damage

The approach to liability based on the duty to negotiate on the reparation of transboundary harm when no internationally wrongful act has been committed has

---

[185] Other examples include Council of Europe 1969 Draft Convention on the Protection of Fresh Water against Pollution, CECA Doc. 2561 (1969) and bilateral 1974 Exchange of Notes Between the United States of America and Canada Constituting an Agreement Relating to Liability for Loss and Damage from Certain Rocket Launches, UNTS, vol. 992, p. 97 (1975).

not taken root in international law due to lack of political acceptance, save a handful of sources with limited applicability.[186] Thus, politically, it is improbable that this approach should be used in the context of climate change-related damage.

While suitable in situations such as those involving a major water project, this approach would be of little utility in the climate change context. The approach based on the obligation to negotiate a redress settlement functions with individual projects, generally prior to their implementation, and, systemically, is unfit to address climate change-related damage, which is caused by GHG emissions from multiple sources.[187]

3.2.1.3 Obligation to Ensure Prompt, Adequate, and Effective Compensation

Another approach to state liability consists in obligating the source state to ensure that victims of transboundary damage receive prompt, adequate, and effective compensation whereby temporal, quantitative, and qualitative considerations are respectively taken into account.

In 1997, the ILC separated the concepts of liability and prevention, and subdivided its work on international liability for injurious consequences arising out of acts not prohibited by international law into two parts: prevention of transboundary damage from hazardous activities and international liability in case of loss from transboundary harm arising out of hazardous activities. In December 2001, the UN General Assembly reviewed, and expressed its appreciation for, the ILC's Draft Articles on Prevention of Transboundary Harm from Hazardous Activities and requested the Commission to proceed with its work on international liability. In 2006, the ILC adopted a set of Principles on Allocation of Loss in the Case of Transboundary Harm Arising Out of Hazardous Activities and later that year, the General Assembly took note of the Principles in its Resolution 61/36 on Allocation of Loss in the Case of Transboundary Harm Arising out of Hazardous Activities.[188]

The obligation to ensure prompt, adequate, and effective compensation performs a reparative function in cases when the state has complied with its due diligence obligation to prevent transboundary environmental harm under customary international law but damage has nonetheless been caused.[189] It is an obligation

---

[186] Lefeber 1996, p. 220.

[187] Political and systemic unsuitability of this approach to climate change-related damage makes considerations of a legal and conceptual nature redundant.

[188] *Allocation of loss in the case of transboundary harm arising out of hazardous activities*, GA Res. 61/36 (4 December 2006).

[189] On the obligation to prevent significant transboundary harm, see Section 5.2.3; see Section 3.2.2.2 for the distinction between absolute obligations and obligations of due diligence.

directed at the achievement of a particular result, one of compensating the victims of transboundary harm in a prompt, adequate, and effective manner. It has been suggested that the obligation to ensure prompt, adequate, and effective compensation should be extended to all transboundary environmental interference, irrespective of its nature, in order to provide victims with financial guarantees against harm caused by hazardous and non-hazardous activities alike.[190] At present, however, this obligation can only be discerned in respect of harm arising out of hazardous activities, which is reflected in the ILC's 2006 Principles on Allocation of Loss in the Case of Transboundary Harm Arising Out of Hazardous Activities. It has been the practice adopted by international agreements and municipal laws to limit the applicability of special civil liability regimes on the basis of the nature of the activity.[191]

As their title indicates, the Principles apply to transboundary damage caused by hazardous activities not prohibited by international law. This notion comprises four elements: (1) such activities are not prohibited by international law; (2) they carry a risk of causing significant harm; (3) harm must be transboundary; and (4) the transboundary harm must be caused by such activities through their physical consequences (harmful impacts caused by trade, monetary, socio-economic or fiscal policies are thereby excluded from the Principles' scope).[192] Thus, the liability regime embodied in the Principles focuses on the consequences of an activity rather than its lawfulness. Under the Principles, hazardous activities are understood to refer to 'activities that have a high probability of causing significant transboundary harm or a low probability of causing disastrous transboundary harm.'[193] Significant harm must be construed to mean 'something more than 'detectable' but not necessarily at the level of 'serious' or 'substantial.'[194] The significance threshold is essential to identify damage eligible for compensation and to ensure that no frivolous claims are launched. Harm that does not reach the level of significant is considered tolerable.[195]

---

[190] Lefeber 1996, p. 233.

[191] See Lefeber 1996, pp. 239-254.

[192] 2006 Principles on the Allocation of Loss in the Case of Transboundary Harm Arising out of Hazardous Activities, ILC Report on the work of its 58th session, A/61/10, forthcoming in YILC, commentary to Principle 1, para. 4 & Principle 2, para. 24.

[193] 2006 Principles on the Allocation of Loss in the Case of Transboundary Harm Arising out of Hazardous Activities, ILC Report on the work of its 58th session, A/61/10, forthcoming in YILC, commentary to Principle 1, para. 2.

[194] 2006 Principles on the Allocation of Loss in the Case of Transboundary Harm Arising out of Hazardous Activities, ILC Report on the work of its 58th session, A/61/10, forthcoming in YILC, commentary to Principle 2, para. 2.

[195] 2006 Principles on the Allocation of Loss in the Case of Transboundary Harm Arising out of Hazardous Activities, ILC Report on the work of its 58th session, A/61/10, forthcoming in YILC, commentary to Principle 2, para. 2.

The Principles provide a general framework for the conclusion of international agreements and adoption of domestic laws to cover situations when the state has complied with its duty of due diligence to prevent transboundary damage but the harm has nevertheless been caused. Non-binding in nature, they encourage states to make all efforts to conclude specific international agreements in respect of particular categories of hazardous activities whereby arrangements should be made with regard to compensation, response measures, and international and domestic remedies (Principle 7). States should also adopt the necessary national measures to implement the Principles through domestic legislation (Principle 8(1)).

The purpose of the Principles on Allocation of Loss in the Case of Transboundary Harm Arising Out of Hazardous Activities is twofold: (a) to ensure prompt and adequate compensation to victims of transboundary damage and (b) to preserve and protect the environment *per se* in the event of transboundary damage, especially with respect to mitigation of damage to the environment and its restoration and reinstatement (Principle 3). The purpose of ensuring prompt and adequate compensation to victims relates to the need to protect victims by requiring 'measures of prevention that as far as possible avoid the risk of loss or injury and, insofar as it is not possible, measures of reparation.' [196] The ILC has treated prevention and reparation of harm as two sides of the same coin as the obligation to ensure compensation to victims derives from situations when, despite due diligence, harm could not be prevented (see Section 3.2.1.2).

The underlying premise of the idea that victims of transboundary damage must be promptly and adequately compensated can be found in the Trail Smelter arbitration and the *Corfu Channel* case and was subsequently reflected in the 1972 Stockholm Declaration (Principle 21) and in the 1992 Rio Declaration (Principle 2):

> States have, in accordance with the Charter of the United Nations and principles of international law, the sovereign right to exploit their own resources pursuant to their own environmental and developmental policies, and the responsibility to ensure that activities within their jurisdiction or control do not cause damage to the environment of other States or of areas beyond the limits of national jurisdiction.

The international community shares a common conviction that states shall cooperate to develop international and municipal law regarding liability and

---

[196] See R.Q. Quentin-Baxter's Schematic Outline, *in* R. Q. Quentin-Baxter, *Third report on international liability for injurious consequences arising out of acts not prohibited by international law*, A/CN.4/360 & Corr.1, YILC, vol. II, Part One (1982), para. 53; Section 5, para. 2; Draft Articles on Prevention of Transboundary Harm from Hazardous Activities, ILC Report on the work of its 53rd session, A/56/10, YILC, vol. II, Part Two (2001).

compensation for victims.[197] The definition of victim is closely connected with the question of standing and covers 'natural and legal persons, and includes the State as custodian of public property.'[198] *Locus standi* is more difficult to establish in instances when harm is caused to the environment *per se, i.e.* to the environment itself irrespective of whether damage is simultaneously caused to persons or property.[199] Thus, some domestic liability regimes provide standing to non-governmental organizations or public trustees.[200]

In giving prominence to the protection of the environment, the Principles recognize the importance of preserving this shared resource for the present and future generations and acknowledge that 'damage to the environment *per se* could constitute damage subject to prompt and adequate compensation, which includes reimbursement of reasonable costs of response and restoration or reinstatement measures undertaken.'[201]

In accordance with their purpose, the Principles adopt a two-pronged approach to liability whereby states are required to ensure prompt and adequate compensation to victims on the one hand, and to take response measures on the other (see Section 3.2.1.4). As regards prompt and adequate compensation, states should take all necessary measures to ensure that it is available to victims of transboundary damage caused by hazardous activities located within their territory or within areas under their jurisdiction or control (Principle 4(1)). Those measures should include the imposition of strict liability on the operator or, where appropriate, other person or entity and such liability should not require proof of fault (Principle 4(2)). States should require operators to establish and maintain financial security for the coverage of compensation claims and, where appropriate, industry-wide funds at the national level should also be established (Principle 4(3) and 4(4)). There are also examples of such funds at the international level, notably the International Oil

---

[197] See Principle 22 of the 1972 Stockholm Declaration on the Human Environment & Principle 13 of the 1992 Rio Declaration on Environment and Development.

[198] 2006 Principles on the Allocation of Loss in the Case of Transboundary Harm Arising out of Hazardous Activities, ILC Report on the work of its 58th session, A/61/10, forthcoming in YILC, commentary to Principle 2, para. 29.

[199] 2006 Principles on the Allocation of Loss in the Case of Transboundary Harm Arising out of Hazardous Activities, ILC Report on the work of its 58th session, A/61/10, forthcoming in YILC, commentary to Principle 2, para. 11.

[200] At the municipal level, the US Oil Pollution Act confers the right to act as a public trustee upon the US Government, a State, an Indian tribe, and a foreign government; further instances include Norway, France, and India (see 2006 Principles on the Allocation of Loss in the Case of Transboundary Harm Arising out of Hazardous Activities, ILC Report on the work of its 58th session, A/61/10, forthcoming in YILC, commentary to Principle 2, para. 30).

[201] 2006 Principles on the Allocation of Loss in the Case of Transboundary Harm Arising out of Hazardous Activities, ILC Report on the work of its 58th session, A/61/10, forthcoming in YILC, commentary to Principle 3, para. 6.

Pollution Compensation Funds (IOPC Funds).[202] According to Kiss & Shelton, the Principles are in line with the existing state practice, which 'largely channels liability to the owner or operator and requires financial guarantees to cover future claims of compensation.'[203] However, should the above-mentioned measures be insufficient to provide adequate compensation, the state of origin should ensure that additional financial resources are made available (Principle 4(5)). Without directly requiring states to set aside funds guaranteeing prompt and adequate compensation to victims, this provision serves as a guideline for the adoption of best practices by the state of origin in order to ensure that 'sufficient financial resources are available in case of damage arising from a hazardous operation situated within its territory or in areas under its jurisdiction.'[204] Further, Principle 6 lists international and domestic measures designed to enable the source state and affected states to ensure minimum standards for the provision of prompt, adequate, and effective compensation to victims. States are expected to ensure those minimum standards by providing their judicial and administrative bodies with the necessary jurisdiction and competence (Principle 6(1)) and granting to victims of transboundary damage access to remedies (Principle 6(2)) and information (Principle 6(5)).

The notion of ensuring prompt and adequate compensation to victims should be seen from the cost internalization perspective, which is at the core of the polluter-pays principle.[205] According to that principle, true economic costs of pollution control, clean-up, and protection measures should stay within the operational costs of the activity itself:

> National authorities should endeavour to promote the internalization of environmental costs and the use of economic instruments, taking into account the approach that the polluter should, in principle, bear the costs of pollution, with due regard to the public interest and without distorting international trade and investment.[206]

---

[202] Information on IOPC Funds is available from: <www.iopcfund.org/>.

[203] Kiss & Shelton 2007a, p. 28; see also 2006 Principles on the Allocation of Loss in the Case of Transboundary Harm Arising out of Hazardous Activities, ILC Report on the work of its 58th session, A/61/10, forthcoming in YILC, Principle 4.

[204] 2006 Principles on the Allocation of Loss in the Case of Transboundary Harm Arising out of Hazardous Activities, ILC Report on the work of its 58th session, A/61/10, forthcoming in YILC, commentary to Principle 4, para. 38.

[205] 2006 Principles on the Allocation of Loss in the Case of Transboundary Harm Arising out of Hazardous Activities, ILC Report on the work of its 58th session, A/61/10, forthcoming in YILC, commentary to Principle 3, para. 11.

[206] 1992 Rio Declaration on Environment and Development, Principle 16.

In seeking to provide operators with an incentive to prevent transboundary harm from hazardous activities, it has often formed the foundation for regimes based on strict liability.[207]

Yet, the application of the polluter-pays principle is not without limitations. Under the Principles, liability is generally channelled towards the operator and does not require proof of fault, the underlying notion being not that operators are always liable but that 'the party with the most effective control of the risk at the time of the accident or has the ability to provide compensation is made primarily liable.'[208] Legal channelling ensures that liability is placed with a single person who is easily identifiable but who is not necessarily the source of the harm.[209] Thus, channelling benefits the victim by simplifying the process of claiming against operators.

The Principles do not support strict liability between states; a state could only be strictly liable if it itself is the operator.[210] It has been noted that in developing the Principles, the ILC changed its focus from state liability to civil liability of operators.[211] Bearing in mind the non-binding nature of the Principles, state obligations *vis-à-vis* non-state activities are limited: only in the event that the measures providing for operator's liability are insufficient should the State of origin make available additional financial resources.[212] The 1997 Convention on Supplementary Compensation for Nuclear Damage aimed at enhancing the system of compensation set up pursuant to the 1963 International Atomic Energy Agency (IAEA) Convention on Civil Liability for Nuclear Damage and the 1963 Brussels Convention supplementary to the 1960 Organisation for Economic Cooperation and Development (OECD) Convention on Third Party Liability in the Field of Nuclear

---

[207] See *e.g.* UNEP Guidelines for the Development of Domestic Legislation on Liability, Response Action and Compensation for Damage Caused by Activities Dangerous to the Environment. Adopted by the UNEP Governing Council in decision SS.XI/5, part B (26 February 2010); 1992 United Nations Economic Commission for Europe (UNECE) Convention on the Protection and Use of Transboundary Watercourses and International Lakes, 31 ILM 1312 (1992) and its 2003 Kiev Protocol on Civil Liability and Compensation for Damage Caused by the Transboundary Effects of Industrial Accidents on Transboundary Waters, Doc. ECE/MP.WAT/11-ECE/CP.TEIA/9 (2003); 1992 UNECE Convention on the Transboundary Effects of Industrial Accidents. 31 ILM 1330 (1992); 1993 Convention on Civil Liability for Damage Resulting from Activities Dangerous to the Environment (Lugano Convention), available from: <http://conventions.coe.int/Treaty/en/Treaties/Html/150.htm>; and Directive 2004/35/EC of the European Parliament and of the Council of 21 April 2004 on environmental liability with regard to the prevention and remedying of environmental damage, OJ 2004 L 143/56.

[208] 2006 Principles on the Allocation of Loss in the Case of Transboundary Harm Arising out of Hazardous Activities, ILC Report on the work of its 58th session, A/61/10, forthcoming in YILC, Principle 4(2); see also commentary to Principle 4, para. 10.

[209] Lefeber 1996, p. 272.

[210] Kiss & Shelton 2007a, p. 28.

[211] Fitzmaurice 2007, p. 1025; see also Lefeber 1996, pp. 197-207, 209-219, especially 217.

[212] 2006 Principles on the Allocation of Loss in the Case of Transboundary Harm Arising out of Hazardous Activities, ILC Report on the work of its 58th session, A/61/10, forthcoming in YILC, Principle 4(5).

Energy provide a good illustration in this regard.[213] The 1997 Convention, for instance, imposes upon the installation state an absolute obligation to ensure the availability of additional funds for the provision of compensation to victims to the extent that the funds from the liable operator are insufficient to cover the first tier amount fixed at 300 million Special Drawing Rights (SDRs).[214] For instances when claims for compensation exceed 300 million SDRs, member states are required to contribute to an international supplementary fund to cover the second tier of compensation, the amount of which is not set. The Convention also provides for the possibility of establishing a third tier of compensation 'provided that damage in a Contracting Party having no nuclear installations within its territory shall not be excluded from such further compensation on any grounds of lack of reciprocity.'[215]

Civil liability regimes governing the use of nuclear power sources have their origin in the colossal amount of harm that can be caused by nuclear accidents and generally channel liability to operators. The 1997 Convention on Supplementary Compensation for Nuclear Damage is noteworthy in this regard as it requires states to set aside additional financial resources to be used in instances when the settlement of claims on the basis of a civil liability regime might be inadequate.

The general lack of focus on state liability (as opposed to liability of operators) in the Principles on Allocation of Loss in the Case of Transboundary Harm Arising Out of Hazardous Activities must be seen in the context of the ILC's draft on Prevention of Transboundary Harm from Hazardous Activities.[216] According to the Draft Articles, non-fulfilment of the due diligence obligation to prevent transboundary harm leads to a breach of an international obligation, which shifts the applicable legal regime to that of state responsibility discussed later in this chapter.[217] If the state has complied with its due diligence duty to prevent harm from a particular hazardous activity but has failed to act in accordance with its obligation to ensure prompt, adequate, and effective compensation for the harm caused, state responsibility may likewise be engaged. In situations when harm has been caused without the commission of an internationally wrongful act, the state, in

---

[213] 1997 IAEA Convention on Supplementary Compensation for Nuclear Damage, INFCIRC/567 (22 July 1998); 1963 IAEA Convention on Civil Liability for Nuclear Damage, UNTS, vol. 1063, p. 265 (1977); 1963 OECD Convention Supplementary to the Convention on Third Party Liability in the Field of Nuclear Energy, UNTS, vol. 1041, p. 358 (1977); 1960 OECD Convention on Third Party Liability in the Field of Nuclear Energy. UNTS, vol. 956, p. 263 (1974).

[214] 1997 IAEA Convention on Supplementary Compensation for Nuclear Damage, INFCIRC/567 (22 July 1998), Art. III(1), not yet in force.

[215] 1997 IAEA Convention on Supplementary Compensation for Nuclear Damage, Art. XII(2).

[216] See Kiss & Shelton 2007a, pp. 27-28.

[217] 2001 Draft Articles on Prevention of Transboundary Harm from Hazardous Activities, ILC Report on the work of its 53rd session, A/56/10, YILC, vol. II, Part Two, commentary to Art. 1, para. 6.

principle, may choose whether to assume state liability or to have claims settled on the basis of a civil liability regime.[218]

3.2.1.3.1   Obligation to Ensure Prompt, Adequate, and Effective Compensation and Climate Change-Related Damage

The principal difference of this approach from the one involving the obligation to pay compensation is that it places primary liability with the operator and not with the state, which directly enables victims to seek compensation for the harm suffered. Albeit climate change-related damage may amount to damage to the environment *per se*, *e.g.* changes in the composition of the atmosphere,[219] it also covers damage of a more conventional nature. Originating from slow onset events (*e.g.* the rising sea levels) or extreme weather events, such damage mainly involves damage to property. If the Principles were to apply to situations of damage caused by anthropogenic climate change, victims would be easily identifiable. Politically and conceptually, an approach to climate change-related damage channelling liability to operators would be feasible but it cannot withstand the rigours of a legal test. On the one hand, channelling liability to operators, *i.e.* GHG emitting entities, could potentially result in unreasonably wide coverage. On the other hand, a potential claimant would run into causation problems due to the multiplicity of GHG emitting sources.

Additionally, the emission of GHGs does not fall under the definition of hazardous activities and the Principles do not obligate states to ensure prompt, adequate, and effective compensation to victims of harm caused by activities that are considered non-hazardous. Thus, the approach to liability based on the obligation to ensure prompt, adequate, and effective compensation cannot be relied upon in addressing climate change-related damage due to systemic incompatibility.

3.2.1.4 Obligation to Take Response Action

The fourth approach to state liability in international law involves the obligation to take response action. The Principles on Allocation of Loss in the Case of Transboundary Harm Arising Out of Hazardous Activities analysed in the previous section provide some general guidance with regard to situations arising after the occurrence of transboundary damage. In accordance with the Principles, upon the

---

[218] Lefeber 1996, pp. 231-234.

[219] Note that 'Protection of the atmosphere' has been included in the ILC Long-Term Programme of Work. See ILC Report on the work of its 63rd session, A/66/10, forthcoming in YILC (2011), p. 7, para. 32.

occurrence of an incident involving a hazardous activity, the state of origin is duty-bound to obtain from the operator the necessary information and promptly notify all (likely to be) affected states (Principle 5(a)). The source state is then expected to ensure that appropriate response measures are taken using the best available technology (Principle 5(b)). It should also consult and cooperate with the affected states to mitigate and, if possible, eliminate the effects of transboundary damage (Principle 5(c)). Once notified, all affected states shall take all feasible measures to mitigate and eliminate the effects of transboundary damage (Principle 5(d)). Additionally, all the states concerned should also seek the assistance of competent international organizations (Principle 5(e)).

The principal examples of international agreements that rely on this approach are Annex VI to the Protocol on Environmental Protection to the Antarctic Treaty regarding liability arising from environmental emergencies (Liability Annex) and the Nagoya-Kuala Lumpur Supplementary Protocol on Liability to the CBD (Supplementary Protocol).[220] In utilising the same approach to liability, Annex VI and the Supplementary Protocol take on somewhat different perspectives, which are set out below.

The Antarctic Treaty System is a unique international legal regime enabling the participating states to cooperate in managing the Antarctic continent. In 1991, the states parties concluded the Protocol on Environmental Protection to the 1959 Antarctic Treaty in order to 'enhance the protection of the Antarctic environment.'[221] The Protocol entered into force in 1998. The Liability Annex to the 1991 Protocol on Environmental Protection was concluded in 2005 and will enter into force once all the twenty-eight Antarctic Treaty Consultative Parties have ratified it.

The liability regime under the Liability Annex is predicated on 'the fact that the operator, having caused an environmental emergency that may have significant and harmful impact on the Antarctic environment, did not take the required response action to avoid or minimize such impact.'[222] In line with the general trend towards civil liability of operators (see previous sections), the state is only liable when it itself is the operator. During negotiations, the discussion was focused on the environmental aspect of liability, the purpose being to protect the environment of the Antarctic. Although the phrase 'environmental damage' does not appear in

---

[220] This approach has also been incorporated into the 2010 UNEP Guidelines for the Development of Domestic Legislation on Liability, Response Action and Compensation for Damage Caused by Activities Dangerous to the Environment.

[221] 1991 Protocol on Environmental Protection to the 1959 Antarctic Treaty, 30 ILM 1455 (1991), preamble.

[222] A. Shibata, How to Design an International Liability Regime for Public Spaces: the Case of the Antarctic Environment, in T. Komori & K. Wellens (Eds.), *Public Interest Rules of International Law: towards Effective Implementation* (347-372), Farnham: Ashgate, 2009, at 352.

the final text, the Liability Annex has 'the comprehensive protection of the Antarctic environment and dependent and associated ecosystems'[223] as its main objective and it has been suggested that it creates a truly environmental liability regime.[224] In accordance with Article 5(1), each Party 'shall require each of its operators to take prompt and effective response action to environmental emergencies,' which is an absolute obligation of conduct insofar as state regulation of operators' conduct is concerned. This obligation can also be qualified as a due diligence obligation of result as the response action itself is not guaranteed. If the operator fails to take such action, the state of that operator as well as other states parties are 'encouraged' to take response action.[225] Parties other than the state of the operator shall not take response action to an environmental emergency unless there is an imminent threat of significant harm to the Antarctic environment.[226] Thus, although the Liability Annex does not establish liability for damage *per se*, it covers both actual as well as imminent environmental impacts in obligating the operator to pay for response action, the scope of which, as has been noted, 'is comparable to "the costs of preventive measures" recognized as one element of environmental damage.'[227]

The Liability Annex defines response action as follows:

> reasonable measures taken after an environmental emergency has occurred to avoid, minimise or contain the impact of that environmental emergency, which to that end may include clean-up in appropriate circumstances, and includes determining the extent of that emergency and its impact.[228]

This definition encompasses clean-up after an environmental emergency has occurred but makes no mention of restoration measures, which rather limits the scope of the Annex.

The true novelty of the regime is in that it provides for the operator's strict liability for the costs of response action that has either been taken by one of the parties or should have been taken but was not.[229] If a non-state operator should

---

[223] 2005 Annex VI to the Protocol on Environmental Protection to the Antarctic Treaty (Liability Arising from Environmental Emergencies), available from: <www.ats.aq/documents/recatt/Att249_e.pdf>, preamble.
[224] Shibata 2009, p. 350.
[225] 2005 Annex VI to the Protocol on Environmental Protection to the Antarctic Treaty (Liability Arising from Environmental Emergencies), Art. 5(2).
[226] 2005 Annex VI to the Protocol on Environmental Protection to the Antarctic Treaty (Liability Arising from Environmental Emergencies), Art. 5(3)(a)-(b).
[227] Shibata 2009, p. 353.
[228] 2005 Annex VI to the Protocol on Environmental Protection to the Antarctic Treaty (Liability Arising from Environmental Emergencies), Art. 2(f).
[229] 2005 Annex VI to the Protocol on Environmental Protection to the Antarctic Treaty (Liability Arising from Environmental Emergencies), Art. 6(1)-(2).

have taken response action but did not and no such action has been taken by any party, the operator is still liable to pay an equivalent of the costs of response action into the Article 12 fund (Art. 6(2)(b)). Along similar lines, in situations when a state operator should have taken response action but did not and neither did any other party, the state operator must pay an equivalent of the costs of response action into the Article 12 fund (Art. 6(2)(a)). Actions for compensation against non-state operators can only be brought within limited time (Art. 7(1)). Operators' liability is limited in amount unless the environmental emergency resulted from an act or omission of the operator, committed with the intent to cause it, or recklessly and with knowledge that such an emergency would probably result (Art. 9). The operator may also invoke exemptions from liability (Art. 8).

In essence, the Liability Annex provides for the operator's liability for environmental emergencies in the Antarctic. The Article 6 distinction between state and non-state operators merely reinforces that notion and Article 10 absolves states of liability for operators other than its state operators:

> A Party shall not be liable for the failure of an operator, other than its State operators, to take response action to the extent that that Party took appropriate measures within its competence, including the adoption of laws and regulations, administrative actions and enforcement measures, to ensure compliance with this Annex.

It has been suggested that the Annex VI liability mechanism 'breaks new ground in international environmental law' as it in effect imposes liability for significant and harmful impact on the environment where there is no economic loss involved.[230] It does so by requiring states to ensure that operators take prompt and effective response action that 'may include clean-up' in environmental emergency situations in Antarctica. Yet, Annex VI stops short of requiring restoration measures to be taken. It is also important to note that the Liability Annex does not cover damage caused to the Antarctic environment by an accumulation of injurious effects of climate change; its scope is limited to environmental emergencies.[231]

The obligation to take response action has also provided the basis for the Nagoya-Kuala Lumpur Supplementary Protocol on Liability and Redress to the CBD adopted in Nagoya in October 2010. However, as has been mentioned, the Supplementary Protocol utilizes a different approach.

---

[230] M. Johnson, 'Liability for environmental damage in Antarctica: the adoption of Annex VI to the Antarctic Environmental Protocol,' 19 *Geo. Int'l Envtl. L. Rev.* 33, 2006, p. 41.

[231] See 2005 Annex VI to the Protocol on Environmental Protection to the Antarctic Treaty (Liability Arising from Environmental Emergencies), Art. 1.

The Supplementary Protocol outlines international rules and procedures in the field of liability and redress for damage resulting from transboundary movements of living modified organisms (LMOs) and will enter into force once forty states have ratified it (Arts. 1, 18(1)). The Supplementary Protocol applies to damage resulting from intentional, unintentional, and illegal transboundary movements of LMOs and obligates states to require operators to inform the competent authority, evaluate damage, and take appropriate response measures (Arts. 3, 5(1)). In turn, the competent authority shall identify the operator which has caused the damage, evaluate the damage, and determine which response measures the operator has to take (Art. 5(2)). The crucial difference of the approach adopted by the Supplementary Protocol from the one set out in the Liability Annex to the 1991 Protocol on Environmental Protection discussed above lies in its extended definition of response measures. The Supplementary Protocol defines response measures as reasonable actions to:

i.   Prevent, minimize, contain, mitigate, or otherwise avoid damage, as appropriate;

ii.   Restore biological diversity through actions to be undertaken in the following order of preference:

    a.   Restoration of biological diversity to the condition that existed before the damage occurred, or its nearest equivalent; and where the competent authority determines this is not possible;

    b.   Restoration by, inter alia, replacing the loss of biological diversity with other components of biological diversity for the same, or for another type of use either at the same or, as appropriate, at an alternative location.[232]

First, it encompasses prevention and minimization of harm in the event of damage caused by an incident involving transboundary movement of LMOs. Second, the Supplementary Protocol takes operators' liability a step further and provides for restoration measures to redress the damage and to eliminate, insofar as it is possible, its consequences by bringing biodiversity to its original state or its nearest equivalent and, in case of loss of biodiversity, its replacement with other components thereof. In accordance with the Supplementary Protocol, the operator shall be required to take response measures if damage has occurred or there is a 'sufficient likelihood' that damage will occur (Art. 5(3)).

Like the Liability Annex, the Supplementary Protocol requires that the costs of response measures be borne by the operator. Should the operator fail to take

---

[232] 2010 Nagoya-Kuala Lumpur Supplementary Protocol on Liability and Redress to the Cartagena Protocol on Biosafety, UNEP/CBD/BS/COP-MOP/5/17 (15 October 2010), Art. 2(2)(d).

response measures, the state's competent authority may implement them (Art. 5(4)). It then has the right to recover from the operator the costs of response measures taken (Art. 5(5)). The operator's liability is subject to the requirement of causation between the damage and the LMO and may be limited financially and in time in accordance with the domestic law of the state concerned (Arts. 4, 7, 8). The operator also has the right to invoke exemptions provided for in domestic law (Art. 6).

3.2.1.4.1   Obligation to Take Response Action and Climate Change-Related Damage

Annex VI to the Protocol on Environmental Protection to the Antarctic Treaty and the Nagoya-Kuala Lumpur Supplementary Protocol on Liability and Redress impose on states an absolute obligation of result to require that operators take appropriate response measures. Unlike international agreements based on compensation-related approaches to state liability (obligation to pay compensation and obligation to ensure prompt, adequate, and effective compensation), these instruments do not focus on compensating the injured party (state or non-state), but rather require states to ensure that operators take response action aimed at the avoidance of damage following an environmental emergency, which may include clean-up (Liability Annex), or going as far as to necessitate restoration measures (Supplementary Protocol).

However, with respect to climate change-related damage, the approach based on the obligation to take response action is a conceptual misfit. This approach is directed at responding to an incident, whereas climate change-related damage is a result of a complex process involving numerous actors across time and space. Also, the obligation to take response measures is mainly directed at the consequences of environmental harm and cannot be relied upon to ensure that states take measures to mitigate climate change. As an adaptation strategy it would likely prove belated and inadequate, particularly in cases of damage associated with slow onset events. Besides, damage that climate change is causing to the environment may not be easy to define, whereas climate change-related damage to property is not covered by the approach necessitating the taking of response measures. Similar to the approach involving the obligation to ensure prompt, adequate, and effective compensation, the duty to take response action is based on the operator's liability, and legal difficulties associated with causation may be impossible to overcome.[233]

---

[233] Conceptual unsuitability of this approach to climate change-related damage makes considerations of a legal, political, and systemic nature redundant.

3.2.1.5 Concluding Remarks on State Liability

As the above analysis has shown, the existing approaches to state liability in international law have inspired a number of international agreements in the field of liability and redress. In making states liable to pay compensation, the Space Liability Convention is a rare example and, generally, in international law there is a strong tilt towards the civil liability of operators. The Space Liability Convention provides for absolute as well as fault-based liability of states. Its application, however, is restricted: it only extends to space activities that at the time the Convention was conceived were exclusively state-operated and no involvement of the private sector appeared likely. Furthermore, under the Space Liability Convention, liability can only arise *vis-à-vis* other states; victims cannot claim compensation directly from the state liable under the Convention. It appears that in international law, the approach to state liability based on the obligation to pay compensation is uncommon and in any case is limited to space activities.

The relatively uncommon approach to liability embodied in the 1997 Convention on the Law of the Non-navigational Uses of International Watercourses focuses on the obligation to negotiate a redress settlement. The difficulty with this approach is that states can neither be required to reach an agreement nor can they be compelled to refer the dispute to a third party for settlement.

Under the Principles on Allocation of Loss in the Case of Transboundary Harm Arising Out of Hazardous Activities, the liability of states is limited to situations when the state itself is the operator whereas the civil liability of operators is strict and victims are provided with direct access to an effective remedy, namely the ability to launch a compensation claim. In requiring states to ensure that operators provide prompt, adequate, and effective compensation to victims of transboundary harm, this approach is rooted in the gravity of damage hazardous activities are likely to cause in case of malfunction. In international law, the requirement to ensure prompt, adequate, and effective compensation is largely confined to regulating liability for the damage caused by activities that are considered hazardous, *e.g.* the use of nuclear power sources[234] or transport of hazardous substances.[235]

The approach to liability adopted by Annex VI to the Protocol on Environmental Protection to the Antarctic Treaty and the Nagoya-Kuala Lumpur

---

[234] *E.g.* 1960 OECD Convention on Third Party Liability in the Field of Nuclear Energy; 1963 IAEA Convention on Civil Liability for Nuclear Damage; 1997 Convention on Supplementary Compensation for Nuclear Damage.

[235] *E.g.* 1969 IMCO International Convention on Civil Liability for Oil Pollution; 1971 Convention Relating to Civil Liability in the Field of Maritime Carriage of Nuclear Material; 1989 UNECE Convention on Civil Liability for Damage Caused During Carriage of Dangerous Goods by Road, Rail and Inland Navigation Vessels, Doc. ECE/TRANS/79 (1989).

Supplementary Protocol on Liability and Redress obligates operators, including state operators, to take response action to redress damage. Response action under the Liability Annex envisages avoidance of damage, including clean-up, while the Supplementary Protocol goes further and provides for restoration measures.

It has been concluded that none of the liability models discussed above can be successfully applied to climate change-related damage. A liability mechanism imposing on states the duty to pay compensation would be an attractive option for countries suffering from the injurious consequences of climate change. However, under international law, this approach is limited to space activities associated with ultrahazardous risks, and in the case of climate change-related damage would be politically unacceptable. Given its nature, the approach based on the obligation to negotiate a redress settlement would hardly be able provide victims with any redress at all and is generally unsuitable due to the fact that multiple activities worldwide contribute to climate change. The obligation to ensure prompt, adequate, and effective compensation would be of use to climate change victims on the ground, *e.g.* members of coastal communities or inhabitants of low-lying islands, but it can only be invoked with regard to hazardous activities and, as such, is systemically not fitted to address climate change-related damage. It is significant that compensation claims analysed in Chapter 4 do not come within the scope of the obligation to ensure prompt adequate, and effective compensation because they have not been brought under domestic laws created pursuant to this obligation. There is nothing in the nature of the emission of GHGs *per se* to suggest that this activity can be qualified as hazardous under international law; GHGs are emitted by virtually every human activity, including farming a field or driving a car, and they only become dangerous when reaching high levels of concentration in the Earth's atmosphere. Finally, liability regimes relying on the duty to take response measures deal with environmental damage whereas, at the end of the day, most climate change-associated harms amount to property damage. It may be argued that desertification, at least to some degree, amounts to damage to the environment *per se* but in any case, subjecting states to the obligation to take response measures would be an instance of too little too late: it is unlikely that the injurious effects of climate change can be avoided and restoration may altogether be impossible. In other words, the approach based on the obligation to take response action cannot accommodate climate change-related damage due to lack of conceptual compatibility.

3.2.2   State Responsibility

During the negotiations of the 1992 UNFCCC, some states advocated including a provision reiterating the rights and obligations of states under the general law of state responsibility. Such a provision did not find its way into the final text of the Convention; however, upon signature, several small island nations made a declaration, which they perceived as necessary to preserve their rights under the law of state responsibility. In the declaration, they stated that the Convention did not prejudice the rules of international law concerning state responsibility and that none of its provisions could be interpreted as derogating from the principles of general international law.[236]

3.2.2.1 State Responsibility: Introduction

State responsibility as reflected in the 2001 ILC Articles on Responsibility of States for Internationally Wrongful Acts is triggered by an internationally wrongful act of a state while conventional law imposing liability on states is considered *lex specialis*. In accordance with the *lex specialis derogat legi generali* maxim, conventional liability provisions take precedence over the general law of state responsibility.[237] In the absence of such provisions, the extraconventional law of state responsibility remains the only available framework for interstate liability and redress. State liability pertains to situations when lawful conduct has resulted in harm, and unless there are such primary rules providing for state liability, a state can only be held liable *ex delicto*, *i.e.* having committed an internationally wrongful act.[238]

The current section takes a careful look at the law of state responsibility and in so doing it generally follows the original structure of the ILC Articles. Section 3.2.2.2 on the origins of state responsibility addresses the concepts of the internationally wrongful act of a state, attributability of wrongful conduct to a particular state, and breach of an international obligation. It also gives consideration to the notion of retroactivity. Next, the various types of international obligations are distinguished: due diligence obligations as opposed to absolute obligations and obligations of

---

[236] See declarations made upon signature by the Governments of Nauru, Tuvalu, Kiribati, Fiji, and Papua New Guinea, available from:
<http://unfccc.int/essential_background/convention/items/5410.php>; P. Sands, *Principles of International Environmental Law* (2nd ed.), Cambridge: Cambridge University Press, 2003, p. 900.
[237] See also 2001 ILC Articles on Responsibility of States for Internationally Wrongful Acts, Art. 55.
[238] Lefeber 1996, p. 55.

conduct *vs.* obligations of result. Further, circumstances precluding wrongfulness are described. Section 3.2.2.3 focuses on the content of state responsibility and the legal consequences of an internationally wrongful act – cessation and reparation. Reparation for injury in general, and restitution, compensation, and satisfaction in particular, are analysed in Section 3.2.2.3.1 where the requirement of causation is given special consideration. Breaches of obligations arising under peremptory norms of general international law are examined in Section 3.2.2.3.2. Finally, Section 3.2.2.4 is devoted to questions of implementation of the international responsibility of states (including countermeasures) and considers, in particular, the issue of standing.

3.2.2.2 Origins of State Responsibility

In international law, states incur responsibility for internationally wrongful acts they commit.[239] State responsibility is engaged when an internationally wrongful act is committed and it (a) is attributable to the state and (b) constitutes a breach of an international obligation (Art. 2). Secondary norms of international law stipulate the origin and consequences of an internationally wrongful act and also govern the implementation of state responsibility.

For there to be an internationally wrongful act of a state, the conduct consisting of an action or omission must first be attributable to that state. In order to determine whether a particular act can be attributed to the state under international law, a distinction must be made between acts of the state and those of private entities operating in the territory of that state or in areas within its jurisdiction or control. Under international law, only the acts or omissions of the state's organs or agents can be attributed to that state; the state cannot be held responsible for the conduct of private actors simply because they may be operating within its jurisdiction or control:

> The General principle drawn from State practice and international judicial decisions, is that the conduct of a person or group of persons not acting on behalf of the State is not considered as an act of the State under international law. This conclusion holds irrespective of the circumstances in which the private person acts and of the interests affected by the person's conduct.[240]

---

[239] 2001 ILC Articles on Responsibility of States for Internationally Wrongful Acts, Art. 1.

[240] 2001 Articles on Responsibility of States for Internationally Wrongful Acts, ILC Report on the work of its 53rd session, A/56/10, YILC, vol. II, Part Two, commentary to Art. 11, p. 119, para. 2; see also Lefeber 1996, p. 56.

In principle, the conduct of any state organ as well as that of bodies authorized to exercise or assume governmental authority is attributable to the state.[241] *Ultra vires* acts of state organs (acts beyond the power granted by law) and acts by private persons directed or controlled by that state are also attributable (Arts. 7, 8). For example, in the *Case Concerning Military and Paramilitary Activities in and Against Nicaragua*, the ICJ attributed responsibility to the US having used the test of effective control over the Contras relating to overflights violating Nicaraguan sovereignty.[242] Conduct that may not have been attributable to the state at the time of commission but which is subsequently adopted by the state as its own is likewise attributable (Art. 11). In its *United States Diplomatic and Consular Staff in Tehran* judgment, the ICJ found that the militants' attack on the Embassy and Consulates could not be considered 'as in itself imputable to the Iranian State.' The Court held, however, that Iran was in violation of its obligation under the Diplomatic and Consular Conventions to protect the premises and that the approval of the militants' attack given by the Iranian state 'translated continuing occupation of the Embassy and detention of the hostages into acts of that State.'[243] In other words, the requirement of attributability is satisfied when an action or omission is carried out by the state – 'a real organized entity, a legal person with full authority to act under international law,'[244] which is without prejudice to the fact that the state 'can only act by and through [its] agents and representatives.'[245]

Lefeber has observed that the non-attribution of private conduct reduces the capacity of international law to protect the environment because the state cannot be held responsible for activities carried out by private persons. Yet, the law of state responsibility does not altogether exclude the possibility of the state being responsible for the injurious consequences of non-governmental activities because '[t]he non-attribution of private conduct is without prejudice to the existence of obligations incumbent on states to regulate and control the conduct of private persons' in their territory and in areas within their jurisdiction or control.[246] Therefore, the state's failure to regulate and control private conduct in accordance with its international obligations, and not the private conduct itself, is attributable to that state under international law and as such has the potential of giving rise to state responsibility. For instance, it has recently been concluded by the ITLOS Seabed

---

[241] See 2001 ILC Articles on Responsibility of States for Internationally Wrongful Acts, Arts. 4-6, 9-10.

[242] *Military and Paramilitary Activities In and Against Nicaragua (Nicaragua v. United States of America)*, Judgment of 27 June 1986, 1986 ICJ Rep. 14, especially para. 86.

[243] *Case Concerning United States Diplomatic and Consular Staff in Tehran (United States of America v. Iran)*, Judgment of 24 May 1980, 1980 ICJ Rep. 3, paras 61-68 and 74.

[244] 2001 Articles on Responsibility of States for Internationally Wrongful Acts, ILC Report on the work of its 53rd session, A/56/10, YILC, vol. II, Part Two, commentary to Art. 2, p. 71, para. 5.

[245] *German Settlers in Poland*, Advisory Opinion, 1923 PCIJ (Ser. B) No. 6, p. 22.

[246] Lefeber 1996, p. 56.

Disputes Chamber in relation to the expression 'to ensure' used in international legal instruments that while it does not imply that the state can be made liable for 'each and every violation committed by persons under its jurisdiction, it is equally not considered satisfactory to rely on mere application of the principle that the conduct of private persons or entities is not attributable to the State under international law.'[247]

Second, in order to establish whether there has been an internationally wrongful act of a state, it is necessary to determine whether the conduct attributable to the state in question constitutes a breach of an international obligation of that state (Art. 2(b)).[248] This is to be done by examining the relevant primary norms of international law as '[t]here is a breach of an international obligation by a state when an act of that state is not in conformity with what is required of it by [the particular] obligation, regardless of its origin or character' (Art. 12). More to the point, if a primary obligation has been violated by conduct attributable to the state, that state has committed an internationally wrongful act.

According to the ILC, conduct prescribed by an international obligation may involve an act or an omission or a combination of the two; it may also involve the passage of legislation or taking of certain administrative or other measures or even a threat of such action. In every such case, if the state's conduct is not in conformity with the conduct legally prescribed by the international obligation, there is a breach of that obligation.[249] International obligations may originate in a treaty, customary international law or a general principle of law; state responsibility may arise from breaches of bilateral obligations, multilateral obligations, or from those owed to the international community as a whole.[250] International obligations may be extended in time or not have a continuing character; they may consist of single or composite acts (Arts. 14, 15). Regardless of the nature or purpose of an international obligation, its breach must occur at the time when the state is bound by it, which offers a guarantee against retroactive application of international law in matters of state responsibility (Art. 13).[251]

---

[247] *Responsibilities and Obligations of States with Respect to Activities in the Area*, Advisory Opinion of 1 February 2011, 2011 ITLOS Rep. 10, para. 112.

[248] Attribution of wrongful conduct to a state in international law is a concept distinctly different from attribution of harm complained of to a particular defendant in domestic law.

[249] 2001 Articles on Responsibility of States for Internationally Wrongful Acts, ILC Report on the work of its 53rd session, A/56/10, YILC, vol. II, Part Two, commentary to Art. 12, p. 125, para. 2.

[250] 2001 Articles on Responsibility of States for Internationally Wrongful Acts, ILC Report on the work of its 53rd session, A/56/10, YILC, vol. II, Part Two, commentary to Art. 12, p. 126, para. 3; p. 128, para. 6.

[251] 2001 Articles on Responsibility of States for Internationally Wrongful Acts, ILC Report on the work of its 53rd session, A/56/10, YILC, vol. II, Part Two, commentary to Art. 13, p. 133, para. 1.

In order to establish whether an international obligation has been breached, it is essential to identify the exact primary norm the state is bound by, its content, and character. The ILC has recognized that some adopted classifications of international obligations may be helpful in ascertaining a breach.[252] The two legal paradigms that may assist in this regard are: obligations of conduct vs. obligations of result and absolute obligations vs. due diligence obligations.[253]

As far as the first binary is concerned, state responsibility is dependent on 'the precise formulation of the duty breached, and in particular whether it requires the state to adopt a specified line of conduct or the achievement of a particular result.'[254] The ILC has noted that 'a distinction is commonly drawn between obligations of conduct and obligations of result' and that '[t]hat distinction may assist in ascertaining when a breach [of an international obligation] has occurred.'[255] This is notwithstanding the fact that the origin or character of an obligation is irrelevant when a state does not act in conformity therewith (Art. 12). Obligations of conduct, or comportment, require a state to adopt a particular course of conduct whereas obligations of result are directed at a particular outcome, irrespective of the means employed towards its achievement. An obligation of conduct is breached 'irrespective of the occurrence of injurious consequences or of any external event;' a state's failure to conduct itself in a way required by an international obligation is sufficient for a finding of breach.[256] Conversely, a state is in breach of an obligation of result when the conduct it has chosen to adopt has not led to the outcome required by the international obligation in question.

The ILC distinction between obligations of conduct and obligations of result has been helpful in classifying international obligations relating to the protection of the environment in order to facilitate the determination of their breach. Such

---

[252] 2001 Articles on Responsibility of States for Internationally Wrongful Acts, ILC Report on the work of its 53rd session, A/56/10, YILC, vol. II, Part Two, commentary to Art. 12, p. 130, para. 11.

[253] For obligations of conduct vs. obligations of result see *Case Concerning the Gabčíkovo-Nagymaos Project (Hungary v. Slovakia)*, Judgment of 25 September 1997, 1997 ICJ Rep. 7, at 77, para. 135; 2001 Articles on Responsibility of States for Internationally Wrongful Acts, ILC Report on the work of its 53rd session, A/56/10, YILC, vol. II, Part Two, commentary to Art. 12, p. 130, para. 11; see also *Responsibilities and Obligations of States with Respect to Activities in the Area*, Advisory Opinion of 1 February 2011, 2011 ITLOS Rep. 10, paras 110-111. For absolute obligations vs. due diligence obligations see *Case Concerning Pulp Mills on the River Uruguay (Argentina v. Uruguay)*, Judgment of 20 April 2010, 2010 ICJ Rep. 14, at 79, para. 197; see also *Responsibilities and Obligations of States with Respect to Activities in the Area*, Advisory Opinion of 1 February 2011, 2011 ITLOS Rep. 10, paras 110-111 and 115-116, and, generally, 2006 Principles on the Allocation of Loss in the Case of Transboundary Harm Arising out of Hazardous Activities.

[254] P. Okowa, *State Responsibility for Transboundary Air Pollution in International Law*, Oxford: Oxford University Press, 2000, p. 171.

[255] 2001 Articles on Responsibility of States for Internationally Wrongful Acts, ILC Report on the work of its 53rd session, A/56/10, YILC, vol. II, Part Two, commentary to Art. 12, p. 130, para. 11.

[256] Lefeber 1996, p. 74.

obligations ultimately aim to prevent transboundary environmental interference causing significant harm, which can be described as a compound obligation, *i.e.* one comprising interrelated but distinct elements and operating within the continuum of prevention and reparation. Thus, the compound obligation requires that the source state take all procedural and substantive measures necessary to avoid, minimize, and repair transboundary environmental interference causing significant harm.[257] Lefeber has suggested addressing those obligations by way of differentiating between (1) procedural obligations, (2) quantitative ecostandards, and (3) qualitative ecostandards.[258] This distinction is instrumental to the analysis of international obligations pertinent to climate change in Chapter 5.

Procedural obligations can be described as obligations of conduct. Procedural obligations include obligations to assess the environmental impact of planned activities; to notify and to consult with the potentially affected states; to negotiate a redress settlement; to monitor and control on-going activities; to safeguard and monitor the sites of abandoned activities; and to ensure prompt, adequate, and effective compensation. Breach of any of these obligations is contingent on whether or not the state has adopted the prescribed course of conduct and does not depend on the achievement of the desired result – that of transboundary harm prevention.

Since quantitative ecostandards presuppose the existence of particular numeric goals to be achieved, they create obligations of result. Quantitative ecostandards notably include obligations to reduce emissions of a certain substance (*e.g.* GHGs) by a certain percentage before a certain time. If particular reduction goals are not met within the allotted time period, the wanted result is not achieved. Therefore, the obligation to reduce emissions would be breached unless it can be inferred that it is a due diligence obligation (see below).

Qualitative ecostandards appear to entail obligations of conduct. Some are relatively straightforward in prescribing requirements of particular conduct, *e.g.* obligation to use best technical means or obligation to use ecologically sound production techniques. Others are less obvious, *e.g.* obligation to take appropriate measures to prevent and abate transboundary environmental interference causing significant harm. According to Lefeber, this obligation must be characterized as that of conduct without prejudice to its ultimate objective of preventing such harm.[259] The obligation to prevent and abate significant transboundary harm refers to the adoption of preventive measures and given the fact that the required course of action may vary in its degree of precision, this obligation must be seen as an obligation of conduct and as such, it is not conditional on the actual occurrence of transboundary environmental harm.

---

[257] Lefeber 1996, p. 34.
[258] Lefeber 1996, pp. 76-77.
[259] Lefeber 1996, pp. 77-79.

The second distinction – that between absolute obligations and due diligence obligations – is based on the state's efforts directed towards compliance. Obligations that require a state to take effective measures 'of a legislative, administrative, or juridical nature to prevent legally protected interests of third states from being harmed by public or private conduct' are due diligence obligations.[260] They must be distinguished from absolute obligations whose breach is independent of the state's efforts to comply with them.

The concept of due diligence has traditionally been used to alleviate the impacts of state responsibility in a number of areas of international law and has proven to be of particular relevance to the regulation of private conduct by the state so that persons within its jurisdiction or control do not harm the interests of third states. Due diligence obligations entail certain minimum standards that the state must observe in its efforts to comply with those obligations. For instance, with regard to hazardous activities involving a risk of significant transboundary harm, states are obliged to act with due diligence in allowing those activities by subjecting them to requirements of prior authorization, environmental impact assessments (EIAs), and subsequent monitoring of environmental impacts. The fact that in cases of transboundary harm arising out of hazardous activities liability attaches to the operator does not release the state from its due diligence duties of prevention.[261] The state's compliance with its customary duty to prevent and abate transboundary environmental interference causing significant harm is measured against its endeavours to comply with it. States are required to take measures towards the prevention of harm; it is irrelevant whether in the end harm is prevented; what is important is that states have to exercise due diligence towards the achievement of this goal. In its Articles on Prevention of Transboundary Harm from Hazardous Activities, the ILC describes this duty as follows:

> The obligation of the State of origin to take preventive or minimization measures is one of due diligence. It is the conduct of the State of origin that will determine whether the State has complied with its obligation [...]. The duty of due diligence involved, however, is not to guarantee that significant harm be totally prevented, if it is not possible to do so. In that eventuality, the State of origin is required [...] to exert its best possible efforts to minimize the risk. In this sense, it does not guarantee that the harm would not occur.[262]

---

[260] Lefeber 1996, p. 61.

[261] 2006 Principles on the Allocation of Loss in the Case of Transboundary Harm Arising out of Hazardous Activities, ILC Report on the work of its 58th session, A/61/10, forthcoming in YILC, general commentary, para. 9.

[262] 2001 Draft Articles on Prevention of Transboundary Harm from Hazardous Activities, ILC Report on the work of its 53rd session, A/56/10, YILC, vol. II, Part Two, commentary to Art. 3, p. 154, para. 7.

In its analysis of the 'responsibility to ensure,' the ITLOS Seabed Disputes
Chamber has captured the essence of a due diligence obligation in the following
manner:

> The sponsoring State's obligation 'to ensure' is not an obligation to achieve, in
> each and every case, the result that the sponsored contractor complies with [its]
> obligations. Rather, it is an obligation to deploy adequate means, to exercise best
> possible efforts, to do the utmost, to obtain this result.[263]

The ITLOS has characterized the obligation to ensure as a due diligence obligation
of conduct noting that in international law, the notions of obligations of conduct
and obligations of due diligence are connected. Shortly before, the ICJ arrived at a
similar conclusion stating that '[a]n obligation to adopt regulatory or administrative
measures [...] and to enforce them is an obligation of conduct' and calling upon the
Parties 'to exercise due diligence in acting through the Commission for the
necessary measures to preserve the ecological balance of the [Uruguay] river.'[264]

Elusive as the exact standard of due diligence may appear to be, an effective
due diligence measure must be likely to lead to the wanted result. In its recent
assessment of a particular treaty obligation, the ICJ has thus described an effective
due diligence measure:

> It is an obligation which entails not only the adoption of appropriate rules and
> measures, but also a certain level of vigilance in their enforcement and the exercise
> of administrative control applicable to public and private operators, such as the
> monitoring of activities undertaken by such operators, to safeguard the rights of the
> other party.[265]

Objectively, the required standard of care should correspond to the degree of due
diligence to be expected of any 'good government' in a comparable situation.[266]
Since states' circumstances are different and some states may have a limited choice
of compliance means at their disposal, the subjective component of the due
diligence standard, too, gains relevance.[267] This is without prejudice to the fact that

---

[263] *Responsibilities and Obligations of States with Respect to Activities in the Area*, Advisory Opinion of
1 February 2011, 2011 ITLOS Rep. 10, para. 110.

[264] *Case Concerning Pulp Mills on the River Uruguay (Argentina v. Uruguay)*, Judgment of 20 April 2010,
2010 ICJ Rep. 14, at 77, para. 187.

[265] *Case Concerning Pulp Mills on the River Uruguay (Argentina v. Uruguay)*, Judgment of 20 April 2010,
2010 ICJ Rep. 14, at 79, para. 197; *Responsibilities and Obligations of States with Respect to Activities in the
Area*, Advisory Opinion of 1 February 2011, 2011 ITLOS Rep. 10, paras 110-116.

[266] 2001 Draft Articles on Prevention of Transboundary Harm from Hazardous Activities, ILC Report on the
work of its 53rd session, A/56/10, YILC, vol. II, Part Two, commentary to Art. 3, p. 155, para. 17.

[267] Lefeber 1996, p. 65.

even states without a well-developed economy, and human and material resources are expected to exercise vigilance and employ their infrastructure.[268] The degree of due diligence to be exercised by the state depends on the means available to that state, the nature of the interests to be protected, and the exact circumstances of the case. Additionally, the degree of due diligence is not a static concept and can change overtime either because 'measures considered sufficiently diligent at a certain moment may become not diligent enough in light, for instance, of new scientific or technological knowledge' or the degree of due diligence 'may [...] change in relation to the risks involved in the activity. [...] [The due diligence standard] has to be more severe for the riskier activities.'[269] For instance, the degree of due diligence associated with the prevention and abatement of transboundary environmental interference causing significant harm has increased since the 1972 UN Conference on the Human Environment due to the changing socio-economic and political values.[270] Lefeber has suggested that the due diligence standard depends on the following: (1) degree of control; (2) object of control; (3) technical and economic capacity of the state concerned; and (4) the interests at stake.[271] To clarify, the state can only be required to exercise due diligence in the territory within its jurisdiction or control (including situations when it functions as the flag state for vessels, aircraft, and spacecraft outside its territory). The standard of due diligence depends on the hazardousness of the regulated activity, the protected interests, and the technical and economic capabilities of states. Insofar as it allows for differentiation in conduct standards for different states, due diligence has been compared to the concept of common but differentiated responsibilities albeit it lacks 'the elements of conditionality and solidarity' of the latter.[272]

In contrast with due diligence obligations, absolute obligations are focused on the desired result rather than action and are phrased in absolute terms. The state's efforts towards the achievement of goals specified by an absolute obligation are of no importance. If the result contemplated by an absolute obligation is not attained, even though the state may have taken steps towards compliance, it would still be in breach, *i.e.* if an international obligation is absolute in character, a state's failure to live up to the goal set by such an obligation 'automatically entails' breach of that obligation. Obligations of conduct, such as those on qualitative ecostandards, are

---

[268] 2001 Draft Articles on Prevention of Transboundary Harm from Hazardous Activities, ILC Report on the work of its 53rd session, A/56/10, YILC, vol. II, Part Two, commentary to Art. 3, p. 154, para. 17.

[269] *Responsibilities and Obligations of States with Respect to Activities in the Area*, Advisory Opinion of 1 February 2011, 2011 ITLOS Rep. 10, para. 117.

[270] Lefeber 1996, pp. 65-66.

[271] Lefeber 1996, pp. 66-69.

[272] Birnie, Boyle & Redgwell 2009, p. 149; see also Chapter 5.

frequently phrased in absolute terms.[273] For example, states may be required to prohibit certain activities, to prescribe the use of particular techniques or to 'take appropriate measures.' The obligation to take appropriate measures is absolute to the extent that it requires the state to *regulate* the conduct of private persons within its jurisdiction or control; however, insofar as it requires the state to *control* private conduct within its territory, the obligation must be described as one of due diligence.[274]

True to their form, absolute obligations are phrased in absolute terms, *e.g.* phrases like 'shall ensure,' 'take effective measures to,' 'shall reduce,' etc. are used.[275] Notably, expressions like 'shall endeavour' or 'shall avoid to the maximum extent possible' are characteristic of due diligence obligations; conditionality is implied despite the absolute character of the modal verb. Finally, it must be mentioned that, in international law, obligations phrased in terms laxer than the unconditional 'shall' (*e.g.* 'should,' 'may,' etc., with the exception of 'may not') are not considered to be of a compulsory character.

Having considered the distinction between obligations of conduct and obligations of result and having compared due diligence obligations with absolute obligations, it is important to reiterate that correct identification of an international obligation across the two binaries may be helpful, and at times is vital, for the determination of breach of that obligation.

In the international law of state responsibility, a limited number of circumstances may preclude the wrongfulness of conduct that would otherwise violate the international obligations of the state concerned. Circumstances precluding wrongfulness do not annul the obligation but rather provide an excuse for temporary non-performance while those circumstances subsist. They largely operate like defences in domestic legal systems and must be distinguished from questions of the jurisdiction of a court or tribunal over a dispute or issues relating to the admissibility of a claim.[276] Circumstances precluding wrongfulness are: consent

---

[273] See *e.g.* Section 3.2.1.4 for analysis of state obligations under the 2005 Liability Annex to the 1991 Protocol on Environmental Protection to the 1959 Antarctic Treaty.

[274] Lefeber 1996, pp. 69-72.

[275] See 1987 Montreal Protocol on Substances that Deplete the Ozone Layer, UNTS, vol. 1522, p. 3 (1997), Art. 2 or 1997 KP, Art. 3(1) ('shall ensure'); 1988 Sofia Protocol to the 1979 Convention on Long-Range Transboundary Air Pollution Concerning the Control of Emissions of Nitrogen Oxides or their Transboundary Fluxes, UNTS, vol. 1593, p. 287 (1998), Art. 2 ('take effective measures to'); 1985 Helsinki Protocol to the 1979 Convention on Long-Range Transboundary Air Pollution on the Reduction of Sulphur Emissions or their Transboundary Fluxes by at least 30 per cent, UNTS, vol. 1480, p. 215 (1997), Art. 2 ('shall reduce').

[276] 2001 Articles on Responsibility of States for Internationally Wrongful Acts, ILC Report on the work of its 53rd session, A/56/10, YILC, vol. II, Part Two, general commentary to Ch. V, p. 169, para. 2 and p. 172, para. 7.

(Art. 20), self-defence (Art. 21), countermeasures (Art. 22), *force majeure* (Art. 23), distress (Art. 24), and necessity (Art. 25).

First, a state's consent to the commission of a particular act by another state precludes that act's wrongfulness to the extent, and for the duration, of the former state's consent. Second, if the act of a state constitutes a lawful measure of self-defence in line with the UN Charter, its wrongfulness is likewise precluded. Third, to the extent that the act of a state constitutes a countermeasure taken against a state responsible for an internationally wrongful act, its wrongfulness is also precluded. Countermeasures are measures that involve the non-performance of international obligations. Countermeasures can only be taken in response to non-performance of an international obligation by another state and only in a way that allows for the resumption of performance of the obligation in question (Arts. 49(2) and 49(3)). A number of fundamental international obligations cannot be affected by countermeasures, namely the obligation to refrain from the threat or use of force, human rights obligations, humanitarian legal obligations prohibiting reprisals, and other obligations under peremptory norms of international law (Art. 50(1)). Countermeasures must be commensurate with the injury suffered and shall be terminated as soon as the responsible state has complied with its international obligations (Arts. 51-53). The fourth circumstance precluding wrongfulness is *force majeure*, *i.e.* the occurrence of an irresistible force or an unforeseen event beyond the control of a state making it materially impossible for that state to perform its obligation (Art. 23). However, *force majeure* does not extend to circumstances that merely make performance of an obligation more difficult. [277] Fifth, distress precludes wrongfulness if the author of the act in question has no other reasonable way, in a situation of distress, of saving the author's life or the lives of those in the author's care. Distress, as well as *force majeure,* cannot be relied upon if it is wholly or in part caused by the state invoking it and if the act in question is likely to cause a comparable or greater peril. The sixth and final circumstance precluding wrongfulness is necessity. The ILC Articles provide a narrow definition of necessity formulated partially in negative terms to highlight the strict limitations safeguarding it against possible abuse. Necessity may not be invoked unless the act in question is the only way for the state to safeguard an essential interest against a grave and imminent peril (Art. 25(1)(a)); it may not be invoked unless the act 'does not seriously impair an essential interest of the State or States towards which the obligation exists, or of the international community as a whole' (Art. 25(1)(b)). Additionally, necessity may not be relied upon as a ground for precluding wrongfulness if the obligation in question excludes such possibility or if the state invoking it has contributed to the situation of necessity (Art. 25). Necessity is

---

[277] 2001 Articles on Responsibility of States for Internationally Wrongful Acts, ILC Report on the work of its 53rd session, A/56/10, YILC, vol. II, Part Two, commentary to Art. 23, p. 184, para. 3.

different from consent, self-defence, and countermeasures in that it is not dependent on the prior conduct of the injured state. It differs from *force majeure* because it does not involve conduct which is involuntary or coerced. Unlike distress that consists in danger to the state official's life or the lives of those in the official's care, necessity is used in situations of grave danger either to an essential interest of the state or the international community as a whole.[278] Although necessity is an available ground to preclude wrongfulness, it is subject to very strict limitations and can only be used in rare cases.

3.2.2.3 Content of State Responsibility

The current section on the content of the international responsibility of states provides insight into the legal relationship the responsible state enters upon commission of an internationally wrongful act. Under the law of state responsibility, an internationally wrongful act of a state entails particular legal consequences (Art. 28). Having committed an internationally wrongful act, the responsible state may not rely on its domestic law as justification for not complying with its international obligations under the law of state responsibility (Art. 32). The core consequences of an internationally wrongful act involve cessation of the wrongful act and full reparation for the injury caused. Reparation may take the form of restitution, compensation and satisfaction (Art. 34). The legal consequences of an internationally wrongful act do not absolve the responsible state from the duty to perform the obligation breached (Art. 29). Even if the state has made full reparation for the injury caused, its continued duty of performance remains unaffected should the breached obligation continue to exist.

Legal consequences of state responsibility include cessation of an internationally wrongful act, which is the primary requirement in eliminating the consequences of wrongful conduct. At the outset, the state responsible for the commission of an internationally wrongful act is obligated to cease that act and offer appropriate assurances and guarantees of non-repetition depending on the specific circumstances of the case (Art. 30). Whether an internationally wrongful act concerns an act or an omission, the function of cessation remains the same: 'to put an end to a violation of international law and to safeguard the continuing validity and effectiveness of the underlying primary rule.'[279]

---

[278] 2001 Articles on Responsibility of States for Internationally Wrongful Acts, ILC Report on the work of its 53rd session, A/56/10, YILC, vol. II, Part Two, commentary to Art. 25, p. 195, para. 2.

[279] 2001 Articles on Responsibility of States for Internationally Wrongful Acts, ILC Report on the work of its 53rd session, A/56/10, YILC, vol. II, Part Two, commentary to Art. 30, p. 218, para. 5.

The second general obligation of the state responsible for an internationally wrongful conduct is closely connected to the concept of cessation and requires the state to make full reparation for the injury caused by its wrongful conduct. Injury in this case may include material or moral damage alike (Art. 31(2)). The essence of the obligation to make reparation is aptly reflected in the Permanent Court of International Justice (PCIJ) *Factory at Chorzów* judgment on the merits:

> The essential principle contained in the actual notion of an illegal act […] is that reparation must, as far as possible, wipe out all the consequences of the illegal act and re-establish the situation which would, in all probability, have existed if the act had not been committed.[280]

In other words, the function of reparation is to re-establish the situation altered by the breach. In its Articles on State Responsibility, the ILC has used the notion of injury in a broad way to include, *inter alia*, harm to the environment resulting from emissions exceeding the prescribed limit although such damage may admittedly be distant, contingent or uncertain.[281]

The obligation to make full reparation arises only if there is a causal link between the internationally wrongful act and the injury. The ILC has referred to a range of terms that have been used to describe this link:

> reference may be made to losses attributable to the wrongful act as a proximate cause, or to damage which is too indirect, remote, and uncertain to be appraised, or to any direct loss, damage including environmental damage and the depletion of natural resources or injury to foreign Governments, nationals and corporations as a result of the wrongful act.[282]

The existence of a causal link between the wrongful act and injury is a necessary condition for reparation. Reparation will not address injury that is too remote or consequential. Among the criteria suitable for defining the causal link, the ILC has listed directness, foreseeability, proximity, whether or not harm was deliberately caused or whether the harm caused was within the ambit of the rule breached.

---

[280] *Factory at Chorzów (Germany v. Poland)*, Merits, Judgment 1928 PCIJ (Ser. A) No.17, p. 47.

[281] 2001 Articles on Responsibility of States for Internationally Wrongful Acts, ILC Report on the work of its 53rd session, A/56/10, YILC, vol. II, Part Two, commentary to Art. 31, p. 226, para. 8.

[282] 2001 Articles on Responsibility of States for Internationally Wrongful Acts, ILC Report on the work of its 53rd session, A/56/10, YILC, vol. II, Part Two, commentary to Art. 31, p. 227, para. 10, footnotes omitted.

3.2.2.3.1   Reparation for Injury

Reparation for the injury caused by an internationally wrongful act can take the form of restitution, compensation, satisfaction or a combination thereof (Art. 34). It is essential to emphasize once again the necessity of the causal link between the internationally wrongful act of the responsible state and the injury suffered by the affected state.

The primary form of reparation is restitution. It is closely related to the concept of cessation of internationally wrongful conduct and in practice the two may sometimes be difficult to distinguish. The main difference lies in the proportionality requirements, which restitution is subject to while cessation is not.[283] A state responsible for an internationally wrongful act is under the obligation to make restitution, *i.e.* to re-establish the situation which existed before the wrongful act was committed provided and to the extent that restitution is not materially impossible and that it does not involve a burden disproportionate to the benefit deriving from restitution instead of compensation (Art. 35). In calling for the re-establishment of the *status quo ante* that existed prior to the commission of the internationally wrongful act 'to the extent that any changes that have occurred in that situation may be traced to that act,'[284] restitution is conditional on the finding of a causal link between the wrongful act and the changes in the situation.

The second form of reparation for injury is compensation. 'It is a well-established rule of international law that an injured State is entitled to obtain compensation from the State which has committed an internationally wrongful act for the damage caused by it' if such damage is not made good by restitution.[285] Whereas the function of restitution is factual re-establishment of the situation that existed prior to the commission of an internationally wrongful act, compensation aims to re-establish the situation legally. The purpose of compensation is to cover any financially assessable damage including loss of profits insofar as it is established (Art. 36(2)). Despite the legal primacy of restitution, compensation is the most frequently sought form of reparation.[286] Compensation focuses on the material damage incurred as a result of an internationally wrongful act and, like any form of

---

[283] 2001 Articles on Responsibility of States for Internationally Wrongful Acts, ILC Report on the work of its 53rd session, A/56/10, YILC, vol. II, Part Two, commentary to Art. 30, p. 218, para. 7.

[284] 2001 Articles on Responsibility of States for Internationally Wrongful Acts, ILC Report on the work of its 53rd session, A/56/10, YILC, vol. II, Part Two, commentary to Art. 35, pp. 237-238, para. 1.

[285] *Case Concerning the Gabčikovo-Nagymaos Project (Hungary v. Slovakia),* Judgment of 25 September 1997, 1997 ICJ Rep. 7, at 81, para. 152.

[286] 2001 Articles on Responsibility of States for Internationally Wrongful Acts, ILC Report on the work of its 53rd session, A/56/10, YILC, vol. II, Part Two, commentary to Art. 36, p. 244, para. 2.

reparation, is dependent on the existence of a causal link between such damage and wrongful conduct preceding it. Compensation is particularly effective in situations when restitution is inadequate or unavailable and at times the injured state would simply prefer compensation to any other form of reparation. The hierarchical relationship between restitution and compensation, including the ways of determining the amount of the latter, has been described by the PCIJ:

> [r]estitution in kind, or, if this is not possible, payment of a sum corresponding to the value which a restitution in kind would bear; the award, if need be, of damages for loss sustained which would not be covered by restitution in kind or payment in place of it – such are the principles which should serve to determine the amount of compensation due for an act contrary to international law.[287]

Compensation has been negotiated by states following attacks on diplomatic missions, injury to embassy personnel, and in cases of damage caused to public property and infrastructure. It is also frequently resorted to outside the state responsibility context; like with the Cosmos 954 satellite incident described earlier in the section on state liability, many of such payments have been made without any admission of responsibility on an *ex gratia* basis.[288] In an environmental context, the Trail Smelter case provides an example whereby the arbitral tribunal awarded compensation for a 'reduction in the value' of the impacted land.[289]

Environmental damage is likewise compensable although, admittedly, it may be difficult to quantify. Environmental damage refers to an adverse or negative effect on the environment that is 'measurable taking into account scientifically established baselines recognized by a public authority' and is significant.[290] Significance of environmental damage may be determined on the basis of factors, such as (a) long-term or permanent change, to be understood as change that will not be redressed through natural recovery within a reasonable period of time; (b) extent of the qualitative or quantitative changes that adversely or negatively affect the environment; (c) reduction or loss of the ability of the environment to provide goods and services, either of a permanent nature or on a temporary basis; (d) extent

---

[287] *Factory at Chorzów (Germany v. Poland)*, Merits, Judgment 1928 PCIJ (Ser. A) No.17, p. 47.

[288] 2001 Articles on Responsibility of States for Internationally Wrongful Acts, ILC Report on the work of its 53rd session, A/56/10, YILC, vol. II, Part Two, commentary to Art. 36, p. 250, para. 12; see also Section 3.2.1.1.

[289] Trail Smelter case (United States v. Canada), 16 April 1938 and 11 March 1941, UNRIAA, vol. III 1905, at 1928.

[290] 2010 UNEP Guidelines for the Development of Domestic Legislation on Liability, Response Action and Compensation for Damage Caused by Activities Dangerous to the Environment, Guideline 3: Definitions, para. 3; see also definition of 'damage' in 2010 Nagoya-Kuala Lumpur Supplementary Protocol on Liability and Redress to the Cartagena Protocol on Biosafety, Art. 2(2)(b).

of any adverse or negative effect/impact on human health; or (e) aesthetic, scientific, and recreational value of parks, wilderness areas, and other lands.[291]

The United Nations Claims Commission (UNCC), responsible for the settlement of claims for damage resulting from Iraq's unlawful invasion and occupation of Kuwait in 1990 and 1991, has awarded compensation for environmental damage in a number of claims. Pursuant to UN Security Council Resolution 687 (1991),

> Iraq [...] is liable under international law for any direct loss, damage, including environmental damage and the depletion of natural resources, or injury to foreign Governments, nationals and corporations, as a result of Iraq's unlawful invasion and occupation of Kuwait.[292]

In the period from 2001 until 2005, the UNCC has dealt with five instalments of claims relating to environmental damage and the depletion of natural resources. Pursuant to Decision No. 7 of the UNCC Governing Council, direct environmental damage and depletion of natural resources include: losses or expenses resulting from reasonable monitoring and assessment of the environmental damage for the purposes of evaluating and abating the harm and restoring the environment; reasonable monitoring of public health and performing of medical screenings for the purposes of investigation and combating increased health risks as a result of environmental damage; losses or expenses resulting from the abatement and prevention of environmental damage, including expenses directly relating to fighting oil fires and stemming the flow of oil in coastal and international waters; reasonable measures to clean up and restore the environment; and the depletion of natural resources.[293] Provided that the causal link to Iraq's invasion and occupation of Kuwait was not merely tenuous, that the chain of causation was uninterrupted, and that the evidentiary challenges were met, the UNCC awarded compensation to the claimant states, albeit in amounts less than originally sought.[294]

---

[291] 2010 UNEP Guidelines for the Development of Domestic Legislation on Liability, Response Action and Compensation for Damage Caused by Activities Dangerous to the Environment, Guideline 3: Definitions, para. 3(b).

[292] SC Res. 687 (8 April 1991), para. 16.

[293] UNCC Governing Council Decision 7. S/AC.26/1991/7/Rev.1 (17 March 1992), para. 35.

[294] See UNCC Governing Council's decisions with respect to the settlement of claims relating to environmental damage and the depletion of natural resources: UNCC Governing Council Decision 132, S/AC.26/Dec.132 (21 June 2001) adopted on the basis of Report and Recommendations Made by the Panel of Commissioners Concerning the First Instalment of 'F4' Claims, S/AC.26/2001/16 (22 June 2001); UNCC Governing Council Decision 171, S/AC.26/Dec.171 (3 October 2002) adopted on the basis of Report and Recommendations Made by the Panel of Commissioners Concerning the Second Instalment of 'F4' Claims, S/AC.26/2002/26 (3 October 2002); UNCC Governing Council Decision 212, S/AC.26/Dec.212 (18 December 2003) adopted on the basis of Report and Recommendations Made by the Panel of Commissioners Concerning the Third Instalment of 'F4' Claims, S/AC.26/2003/31 (18 December 2003);

Although the ILC has admitted that oftentimes environmental damage goes beyond harm that is easily quantifiable (see the definition of 'environmental damage' above), *e.g.* damage to biological diversity, such 'non-use values' are nonetheless real and compensable. [295] Thus, difficulties associated with the quantification of damage do not, in and of themselves, excuse the responsible state from the obligation to compensate the injured state insofar as full reparation is not achieved by restitution.[296]

The third form of reparation is satisfaction. While the other two forms of reparation take precedence, the responsible state is under an obligation to give satisfaction for the injury caused insofar as it cannot be made good by restitution or compensation. Like restitution and compensation, satisfaction is subject to the requirement of causation. Satisfaction may consist in an acknowledgement of the breach, an expression of regret, a formal apology or another appropriate modality and shall not be disproportionate to the injury (Art. 37). Unlike compensation, satisfaction is meant to address non-material injury and is known for its 'rather exceptional character' as it provides a remedy for injuries that cannot be assessed financially.[297] Those injuries have been said to be 'of a symbolic character, arising from the very fact of the breach of the obligation, irrespective of its material consequences for the State concerned.'[298]

Article 37 does not provide an exhaustive list of satisfaction modalities; the appropriate form of satisfaction depends on the circumstances of the particular case.[299] Assurances and guarantees of non-repetition may amount to a form of satisfaction. When the state has suffered a non-material or moral injury, a declaration of wrongfulness of the act in question can be made by a competent

---

UNCC Governing Council Decision 234, S/AC.26/Dec.234 (9 December 2004) adopted on the basis of Report and Recommendations Made by the Panel of Commissioners Concerning Part One of the Fourth Instalment of 'F4' Claims, S/AC.26/2004/16 (9 December 2004); UNCC Governing Council Decision 235, S/AC.26/Dec.235 (9 December 2004) adopted on the basis of Report and Recommendations Made by the Panel of Commissioners Concerning Part Two of the Fourth Instalment of 'F4' Claims, S/AC.26/2004/17 (9 December 2004); UNCC Governing Council Decision 248, S/AC.26/Dec.248 (30 June 2005) adopted on the basis of Report and Recommendations Made by the Panel of Commissioners Concerning the Fifth Instalment of 'F4' Claims, S/AC.26/2005/10 (30 June 2005.

[295] 2001 Articles on Responsibility of States for Internationally Wrongful Acts, ILC Report on the work of its 53rd session, A/56/10, YILC, vol. II, Part Two, commentary to Art. 36, pp. 251-252, para. 15.

[296] More generally, on compensation for environmental damage, see T. Hardman Reis, *Compensation for Environmental Damages under International Law: the Role of the International Judge*, Alphen aan den Rijn: Kluwer Law International, 2011.

[297] 2001 Articles on Responsibility of States for Internationally Wrongful Acts, ILC Report on the work of its 53rd session, A/56/10, YILC, vol. II, Part Two, commentary to Art. 37, p. 263, para. 1.

[298] 2001 Articles on Responsibility of States for Internationally Wrongful Acts, ILC Report on the work of its 53rd session, A/56/10, YILC, vol. II, Part Two, commentary to Art. 37, p. 264, para. 3.

[299] 2001 Articles on Responsibility of States for Internationally Wrongful Acts, ILC Report on the work of its 53rd session, A/56/10, YILC, vol. II, Part Two, commentary to Art. 37, p. 266, para. 5; see also Rainbow Warrior (New Zealand v. France), 30 April 1990, UNRIAA, vol. XX, p. 215 (2006).

court; in fact, declaratory relief is one of the most common forms of satisfaction. The ICJ made a declaration of wrongfulness of the actions of the British Navy in its landmark *Corfu Channel* judgment, which was deemed to provide reparation in the form of satisfaction. The Court ruled that 'to ensure respect of international law, of which it is the organ, the Court must declare that the action of the British Navy constituted a violation of Albanian sovereignty.'[300] The Court made this declaration as per Albania's request deeming it appropriate satisfaction in itself. In the operative part of the judgment, the Court unanimously stated:

> The Court [...] [g]ives judgment that by reason of the acts of the British Navy in Albanian waters in the course of the Operation of November 12th and 13th, 1946, the United Kingdom violated the sovereignty of the People's Republic of Albania, and that *this declaration by the Court constitutes in itself appropriate satisfaction.*[301]

Notably, while the *Corfu Channel* declaration was in itself a form of satisfaction and the only remedy obtained, any competent court would normally determine the legality of a state's act and make a preliminary declaration of lawfulness that may or may not be a form of reparation.

3.2.2.3.2 Serious Breaches of Obligations under Peremptory Norms of International Law

A peremptory norm of international law is one which is 'accepted and recognized by the international community of States as a whole as a norm from which no derogation is permitted and which can be modified only by a subsequent norm of general international law having the same character.'[302] In its identification of peremptory norms, the 1969 Vienna Convention on the Law of Treaties requires that (1) such a norm should meet all of the criteria for recognition as a norm of general international law and (2) that it should be recognized as a norm from which no derogation is permitted by the international community as a whole. Albeit fairly few peremptory norms have been recognized by various national and international tribunals, the concept of *ius cogens* is not confined to the context of treaty validity.[303] The

---

[300] *Corfu Channel (United Kingdom of Great Britain and Northern Ireland v. Albania)*, Judgment of 9 April 1949, ICJ Rep.4, at 35.

[301] *Corfu Channel (United Kingdom of Great Britain and Northern Ireland v. Albania)*, Judgment of 9 April 1949, ICJ Rep.4, at 36, emphasis added.

[302] 1969 Vienna Convention on the Law of Treaties, UNTS, vol. 1155, p. 331 (1987), Art. 53.

[303] 2001 Articles on Responsibility of States for Internationally Wrongful Acts, ILC Report on the work of its 53rd session, A/56/10, YILC, vol. II, Part Two, commentary to Art. 26, p. 208, para. 5.

peremptory norms of general international law that enjoy clear recognition by the international community of states include the prohibitions of slavery, aggression, genocide, torture, racial discrimination, and crimes against humanity as well as the right to self-determination.[304] This list, however, is not exhaustive as new peremptory norms may come into existence through the process of their acceptance and recognition by the international community as a whole.[305]

Under the law of state responsibility, serious breaches of obligations arising under peremptory norms of general international law entail particular consequences specific for such breaches. Should such a breach occur, states are required to cooperate to bring it to an end through lawful means and no state shall render assistance in maintaining a situation of breach. Consequences resulting from serious breaches of obligations arising under peremptory norms of international law are without prejudice to legal consequences of an internationally wrongful act described earlier (Arts. 40 and 41).

3.2.2.4 Implementation of International Responsibility of States

International responsibility of a state arises independently of its invocation by another state; it is only conditional on the occurrence of an internationally wrongful act which, in turn, rests on the two prongs of attributability and breach of an international obligation.[306] Under the law of state responsibility, an internationally wrongful act of a state gives rise to the secondary obligations of cessation and reparation. The ILC provisions on implementation of state responsibility specify what action may be taken by other states to ensure that the responsible state complies with its obligations of cessation and reparation. Implementation of state responsibility determines what states have standing to invoke the international responsibility of the responsible state and what modalities such invocation may be expressed in. The invocation of state responsibility is based on the notion of the injured state – 'the State whose individual right has been denied or impaired by the internationally wrongful act or which has otherwise been particularly affected by that act.'[307] However, the ILC Articles recognize that there may be a broader range of states having a legal interest in invoking state responsibility and securing the

---

[304] 2001 Articles on Responsibility of States for Internationally Wrongful Acts, ILC Report on the work of its 53rd session, A/56/10, YILC, vol. II, Part Two, commentary to Art. 26, p. 208, para. 5.

[305] 2001 Articles on Responsibility of States for Internationally Wrongful Acts, ILC Report on the work of its 53rd session, A/56/10, YILC, vol. II, Part Two, commentary to Art. 40, p. 284, para. 6.

[306] 2001 ILC Articles on Responsibility of States for Internationally Wrongful Acts, Arts. 1 and 2; see also Section 3.2.2.2.

[307] 2001 Articles on Responsibility of States for Internationally Wrongful Acts, ILC Report on the work of its 53rd session, A/56/10, YILC, vol. II, Part Two, general commentary to Part Three, Ch. I, p. 293, para. 2.

performance of the obligation in question. Yet, in principle, the right of invocation of state responsibility is limited to situations when such a right is specifically conferred by a treaty or, in the alternative, the state invoking it must be considered an injured state or be specially affected by the breach. First and foremost, an injured state can invoke the responsibility of another state if the obligation breached is owed to that state individually. It can also invoke the responsibility if the obligation in question is owed to a group of states including that state or to the international community as a whole if the breach (a) specially affects that state or (b) 'is of such a character as radically to change the position of all the other States to which the obligation is owed with respect to the further performance of the obligation' (Art. 42).

Obligations owed to an injured state individually can be described as individual obligations. Obligations that are not owed to a particular state individually but rather apply between more than two states are collective obligations. Examples of obligations owed to an injured state individually include obligations under bilateral treaties and obligations under multilateral treaties or customary international law owed to one particular state. As opposed to individual obligations, collective obligations may be owed to a group of states or to the entire international community, and to invoke responsibility, an injured state would have to be specially affected by the violation of such an obligation. An example of a violation of an international obligation owed to a group of states but specially affecting one or a small number of states is pollution of the high seas in contravention of the 1982 UN Convention on the Law of the Sea (UNCLOS).[308] The UNCLOS requires states to take measures to prevent, reduce, and control pollution of the marine environment. As a result of breach of this obligation, a limited number of states may be particularly affected in that their beaches may be polluted or their coastal fisheries may be harmed by the toxic residues. Those states would then be considered injured and would be entitled to invoke state responsibility notwithstanding the fact that all states parties to the UNCLOS have an interest in the preservation of the marine environment. Collective obligations, the breach of which affects every state to which the obligation is owed, constitute a special category of obligations. In such situations, all states are individually entitled to react to a breach and invoke responsibility. For example, all states participating in the Antarctic Treaty would be injured if one state party were to claim sovereignty over an unclaimed area in Antarctica and as such, they would all be entitled to

---

[308] Pollution of the high seas is in violation of Art. 194 of the 1982 UN Convention on the Law of the Sea. UNTS, vol. 1833, p. 3 (1998); see also 2001 Articles on Responsibility of States for Internationally Wrongful Acts, ILC Report on the work of its 53rd session, A/56/10, YILC, vol. II, Part Two, commentary to Art. 42, pp. 299-300, para. 12.

invoke state responsibility and demand cessation and reparation.[309] Consequently, obligations arising from the commission of an internationally wrongful act, *i.e.* those of cessation and reparation, may likewise be owed by the responsible state to another state, to several states or, in principle, to the international community as a whole (Arts. 32, 33).

In order for the injured state to invoke the responsibility of another state, it must notify the responsible state of its claim. In its claim, the injured state may specify what action the responsible state should take to cease its wrongful conduct. It may also elect the form of reparation it would consider suitable to remedy the situation in question (Art. 43). Any claim must be brought in accordance with the nationality of claims rule and local remedies must be exhausted if the claim in question is subject to such rules (Art. 44). However, if the local remedy does not offer any possibility of redress, for instance when it is apparent that the laws the local court would apply would lead to the rejection of any appeal, the exhaustion of local remedies ceases to be a requirement; only 'available and effective' local remedies have to be exhausted. It has also been suggested that local remedies do not need to be exhausted where there was no relevant connection between the injured person and the State alleged to be responsible at the date of the injury.[310] This exception applies to situations when local remedies are available and might even be potentially effective but 'it would be unreasonable or cause great hardship to the injured alien to exhaust local remedies.'[311] Arguably, where there is no voluntary link or territorial connection between the injured individual and the responsible state, such as in the cases of transboundary environmental harm, there is no need to exhaust local remedies.[312] This has not yet crystallized into a clear rule; however, cases, in which local remedies were 'dispensed with' in the absence of a voluntary link, afford support to the exception to the local remedies rule.[313]

The invocation of responsibility by the injured state is also subject to some exceptions. The injured state may not invoke the responsibility of another state (a) if it has waived the breach itself or its consequences or (b) where it can be inferred from its conduct that the injured state has validly acquiesced in the lapse of the claim (Art. 45).

---

[309] In accordance with Art. IV(2) of the 1959 Antarctic Treaty, UNTS, vol. 402, p. 71 (1961), no new claim to territorial sovereignty in Antarctica shall be asserted while the Treaty is in force; see also 2001 Articles on Responsibility of States for Internationally Wrongful Acts, ILC Report on the work of its 53rd session, A/56/10, YILC, vol. II, Part Two, commentary to Art. 42, p. 300, para. 14.

[310] Diplomatic Protection, GA Res. 62/67 (8 January 2008), Art. 15(c).

[311] 2006 Articles on Diplomatic Protection, ILC Report on the work of its 58th session, A/61/10, forthcoming in YILC, commentary to Art. 15(c), para. 7.

[312] 2006 Articles on Diplomatic Protection, ILC Report on the work of its 58th session, A/61/10, forthcoming in YILC, commentary to Art. 15(c), paras 7-8.

[313] Trail Smelter case (United States *v.* Canada), 16 April 1938 and 11 March 1941, UNRIAA, vol. III 1905; see also 2006 Articles on Diplomatic Protection, ILC Report on the work of its 58th session, A/61/10, forthcoming in YILC, commentary to Art. 15(c), para. 9.

If several states are injured by the same internationally wrongful act, each one of those states may separately invoke the responsibility of the state which has committed the wrongful act (Art. 46). In the ILC Articles on State Responsibility, the possibility of plurality of injured states is matched by the possibility of plurality of responsible states: in situations where several states are responsible for the same internationally wrongful act, the responsibility of each one of those states may be invoked (Art. 47). According to the principle of independent responsibility upon which the ILC's Articles are predicated, in cases of a plurality of responsible states, each state is separately responsible for the conduct attributable to it and its responsibility is not diminished or reduced by the fact that other states are responsible for the same act. State responsibility is determined on the basis of an internationally wrongful act contingent on the breach of a primary norm of international law and 'the responsibility of each [...] State is determined individually, on the basis of its own conduct and by reference to its own international obligations.'[314]

While it is true that the state invoking responsibility is usually the injured state, there are particular instances when a state other than the injured state is also entitled to invoke it. It may do so if (1) the obligation breached is owed to a group of states including that state, and is established for the protection of a collective interest of the group; or (2) the obligation breached is owed to the international community as a whole (Art. 48). The first situation concerns collective obligations subject to a further requirement of a collective interest. Such obligations may relate to the environment or security of the region; they may derive from multilateral treaties or customary international law and have sometimes been referred to as 'obligations *erga omnes partes.*'[315]

The second situation refers to obligations that are owed to the international community as a whole and all states have a legal interest in compliance. In its *Barcelona Traction* judgment, the ICJ has characterized such obligations as follows:

> An essential distinction should be drawn between the obligations of a State towards the international community as a whole, and those arising vis-à-vis another State in the field of diplomatic protection. By their very nature the former are the concern of all States. In view of the importance of the rights involved, all states can be held to have a legal interest in their protection; they are obligations erga omnes.[316]

---

[314] 2001 Articles on Responsibility of States for Internationally Wrongful Acts, ILC Report on the work of its 53rd session, A/56/10, YILC, vol. II, Part Two, commentary to Art. 47, p. 318, para. 8.

[315] 2001 Articles on Responsibility of States for Internationally Wrongful Acts, ILC Report on the work of its 53rd session, A/56/10, YILC, vol. II, Part Two, commentary to Art. 48, p. 320, para. 6.

[316] *Case Concerning the Barcelona Traction, Light and Power Company, Limited (Belgium v. Spain),* Judgment of 5 February 1970, ICJ Rep. 3, at 32, para. 33.

Although no *erga omnes* obligation was at stake in the *Barcelona Traction* case, the above statement indicates that in the field of state responsibility, certain obligations are owed to the international community as a whole and that all states have a legal interest in the protection of such obligations.[317] Each state by virtue of being a member of the international community is entitled to invoke the responsibility of another state for breaches of obligations *erga omnes*.

In practice, it is often the case that obligations under peremptory norms of general international law and obligations *erga omnes* overlap. Yet, while the former focus on the scope and priority to be given to certain fundamental obligations, the latter focus on the legal interest of all states in compliance. Although all peremptory norms are obligations *erga omnes*, the reverse is not true. In this regard, the ILC has made two important points. Firstly, serious breaches of obligations arising under peremptory norms of international law can generate additional consequences not only for the responsible state but for the international community as a whole (see Section 3.2.2.2). Secondly, all states can invoke responsibility for breaches of obligations owed to the international community as a whole.[318]

To recap, in order to have standing to invoke the international responsibility of another state, the invoking state must establish injury. There are situations when a state other than the injured state also has the right to invoke state responsibility. It can do so when the obligation breached is established for the protection of a collective interest of, and is owed to, a group of states including that state; and when the obligation breached is owed to the international community as a whole. States invoking responsibility under Article 48 may claim a declaratory remedy and cessation. Reparation may also be sought in the interest of the injured state or of other beneficiaries of the obligation breached. Thus, for breaches of obligations owed to the international community as a whole, all states have the right to invoke state responsibility and may claim cessation of the internationally wrongful act as well as reparation.[319]

3.2.2.4.1   Countermeasures

Countermeasures are measures taken by an injured state that, if not taken in response to an internationally wrongful act to obtain cessation and reparation,

---

[317] 2001 Articles on Responsibility of States for Internationally Wrongful Acts, ILC Report on the work of its 53rd session, A/56/10, YILC, vol. II, Part Two, general commentary to Part Two, Ch. III, p. 278, para. 2.

[318] 2001 Articles on Responsibility of States for Internationally Wrongful Acts, ILC Report on the work of its 53rd session, A/56/10, YILC, vol. II, Part Two, general commentary to Part Two, Ch. III, pp. 281-282, para. 7.

[319] On the right to claim compensation for breaches of obligations *erga omnes* see *Responsibilities and Obligations of States with Respect to Activities in the Area*, Advisory Opinion of 1 February 2011, 2011 ITLOS Rep. 10, para. 180.

would be considered contrary to the international obligations of that state. With
respect to the origins of state responsibility, the ILC lists countermeasures as one of
the circumstances precluding wrongfulness (Art. 22; see also Section 3.2.2.2).
Countermeasures are subject to a number of important conditions and limitations
because, as unilateral measures, they must be protected against possible abuse. First,
countermeasures must be 'non-forcible' and must not deviate from certain basic
obligations, *i.e.* they shall not affect the obligation to refrain from the use of force;
obligations protecting the fundamental human rights; obligations prohibiting
reprisals; and other obligations under peremptory norms of international law (Art.
50(1)).[320] Second, countermeasures may only be taken against the responsible state
and may not be directed at third parties (Arts. 49(1), 49(2)). Third, an injured state
may only take countermeasures against the state responsible for an internationally
wrongful act in order to induce it to comply with its international obligations, *i.e.*
countermeasures may not be resorted to in order to punish the responsible state
(Arts. 49(1), 51). Fourth, countermeasures are temporary and must be terminated as
soon as the responsible state has achieved compliance (Arts. 49(2), 49(3) and 53).
Fifth, they shall, as far as possible, be taken in a way that would permit the
responsible state to resume the performance of the international obligation breached
(Art. 49(3)). Sixth, countermeasures must, to the extent possible, be reversible in
their effects (Arts. 49(2), 49(3), and 53). Seventh, countermeasures are subject to
the requirement of proportionality (Art. 51). Countermeasures that are not
commensurate with the injury suffered are not lawful and can give rise to state
responsibility. For example, in the *Gabčíkovo-Nagymaros Project* case, the ICJ has
extensively addressed the question of proportionality of countermeasures and ruled
that 'the diversion of the Danube carried out by Czechoslovakia was not a lawful
countermeasure because it was not proportionate' as it deprived Hungary of its
right to an equitable and reasonable share of the Danube's natural resources.[321]

Procedurally, when resorting to countermeasures, an injured state must fulfil a
number of conditions: it shall call upon the responsible state to fulfil its obligations,
notify it of any decision to take countermeasures, and offer to negotiate (Art. 52(1)).
If an internationally wrongful act has ceased or if the dispute has been submitted to a
competent court or tribunal and the dispute settlement procedure is being implemented
in good faith, the taking of countermeasures is unjustified (Arts. 52(3), 53(4)).

Although there have been some instances when states have taken economic
sanctions and other measures with respect to alleged breached of obligations

---

[320] See also 2001 Articles on Responsibility of States for Internationally Wrongful Acts, ILC Report on the
work of its 53rd session, A/56/10, YILC, vol. II, Part Two, general commentary to Ch. II (Countermeasures),
p. 327, para. 6.
[321] *Case Concerning the Gabčíkovo-Nagymaos Project (Hungary v. Slovakia),* Judgment of 25 September
1997, 1997 ICJ Rep. 7, at 56, paras 85, 87.

referred to in Article 48 without claiming to be injured,[322] in principle, states that have standing to invoke responsibility under Article 48 may not take countermeasures in the collective interest.[323]

Alternatively, economic sanctions may be justified as retortions – unfriendly, but otherwise lawful acts, – which would only be in violation of international law if the state is bound by a treaty prohibiting such actions.

3.3 Concluding Remarks

In this chapter, the fundamental difference between state responsibility and state liability has been discussed and the various approaches to the latter have been singled out. The distinctions made point to a certain degree of fragmentation in the field of interstate liability. Whereas it has been concluded that state liability is of little utility in the climate change context, a detailed overview of the law of state responsibility appears to warrant further analysis. In order to evaluate whether the law of state responsibility can provide a legal framework for the accountability of states for climate change-related damage, Chapter 5 identifies primary obligations relating to climate change and analyses the practical application of the secondary rules of state responsibility in the international climate policy context. Prior to that, given the absence of international jurisprudence on interstate liability for climate change-related damage, Chapter 4 examines a range of climate change legal actions launched in various national jurisdictions in order to provide a basis for the identification of general principles of law. In so doing, Chapter 4 identifies and analyses the legal hurdles associated with the climate change phenomenon owing to its complexity and relatively limited exposure to adjudicative scrutiny.

---

[322] *Inter alia,* the ILC mentions the following examples: *United States-Uganda (1978)* when the US adopted legislation prohibiting exports and imports from Uganda to dissociate itself from the latter due to alleged genocide carried out by the Ugandan government; *Certain Western Countries-Poland and the Soviet Union (1981)* when Aeroflot's and LOT's landing rights were suspended by the US, Great Britain, France, the Netherlands, Switzerland, and Austria in response to Poland's imposition of martial law, suppression of demonstrations, and subsequent detention of dissidents; and *Collective measures against the Federal Republic of Yugoslavia (1998)* when in response to the humanitarian crisis in Kosovo the Member States of the European Community adopted legislation freezing the Yugoslav funds and providing for an immediate flight ban.

[323] 2001 Articles on Responsibility of States for Internationally Wrongful Acts, ILC Report on the work of its 53rd session, A/56/10, YILC, vol. II, Part Two, commentary to Art. 54, p. 355, para. 6; see also Fitzmaurice 2007, p. 1021.

4   Liability for Climate Change-Related Damage in Domestic Courts

4.1 Introduction

The past decade has seen a rise in climate change litigation worldwide. The number of climate change-related claims filed in 2010 in the US alone has reached as many as 170.[324] Just a few years ago national courts took up only a small number of climate change cases. By 2013, terms like 'challenges to state action slide,' 'challenges to federal action slide,' and 'Clean Air Act slide' have become part of US lawyers' vocabulary. Climate change litigation has also been spreading to other jurisdictions.[325] However, until today, no GHG emitter has been found liable for climate change-related damage by any domestic court.

Climate change lawsuits, albeit geographically widespread, have been limited to a handful of jurisdictions, predominantly common law systems, with the US and Australia leading the way. This chapter discusses climate change litigation in domestic courts and identifies the legal challenges to establishing liability for climate change-related damage. In its assessment of climate change jurisprudence, the chapter draws mainly on the case law from the US and Australia because those two countries have seen the greatest share of climate change claims litigated to date. It is submitted that domestic litigation insofar as liability principles are concerned may inform international liability principles through the concept of general principles of law.[326] It must be acknowledged that there exist considerable challenges to transposing domestic principles to international law. A general principle of law is a proposition of law so fundamental that it will be found in virtually every legal system and can usually be identified through a comparative law process.[327] It is admitted that, albeit on the rise, domestic litigation of climate change-related claims has not yet permeated many legal systems. At the moment, it is debatable whether the overall number of adjudicated claims in a limited number of jurisdictions can constitute a representative sample to warrant a comparative

---

[324] Columbia Center for Climate Change Law; <www.climatecasechart.com/> (last visited on 11 January 2013).

[325] *E.g.* Climate change-related actions have been launched in the US, Australia, New Zealand, the UK, Canada, the European Union, Germany, France, the Philippines, Nigeria, Ukraine, and the Czech Republic; see Columbia Center for Climate Change Law; <www.climatecasechart.com/> (last visited on 11 January 2013).

[326] M.G. Faure & A. Nollkaemper, 'International liability as an instrument to prevent and compensate for climate change,' 43A *Stan. J. Int'l L.* 123, 2007, p. 129.

[327] J.F. Murphy, *The Evolving Dimensions of International Law: Hard Choices for the World Community*, New York: Cambridge University Press, 2010, p. 25.

approach. The present author has deemed it instructive to consider the existing climate change-related claims, fully aware of the above-mentioned difficulties. A comparative study of climate litigation may merit further scholarly consideration in the future.

Climate change plaintiffs, ranging from environmental groups to federal states and private individuals, have brought actions in domestic courts inculpating corporations, government agencies, oil refineries, motor vehicle manufacturers, power plants, and other public and private entities. *Inter alia,* actions have been brought in tort (*e.g.* public nuisance, negligence, civil conspiracy, misrepresentation), under administrative law (including merits review and judicial review), and constitutional law. One goal of climate change litigation has been 'to impose legal liability upon a party that is somehow responsible for the emission of greenhouse gases that contribute to climate change,'[328] albeit in practice, climate change lawsuits have targeted a broader range of issues, such as forcing municipal or federal governments to act or challenging the approval of particular GHG-intensive projects or projects likely to be affected by climate change impacts.

Since most of the cases have been adjudicated in US and Australian courts, the growing public concern with climate change appears to be reinforced by relative non-involvement of the legislative and executive branches of government and the slow pace of regulation on the matter. It must be noted that although Australia has ratified the Kyoto Protocol,[329] it remains one of the largest *per capita* emitters in the world due to its reliance on coal-fired power production. The US has not ratified the Kyoto Protocol and therefore, will not be bound by any emissions reduction commitments under it. To reduce its carbon output, the US has elected to tackle the global climate change challenge through domestic action, which is not based on any international treaty.

The perceived lack of government action has prompted environmental groups and other plaintiffs to resort to the help of the judiciary with a view to fill legislative lacunae with judicial precedent.[330] This upsurge of climate change

---

[328] S.-L. Hsu, 'A realistic evaluation of climate change litigation through the lens of a hypothetical lawsuit,' 79 *U. Colo. L. Rev.* 701, 2008, p. 702.

[329] The Australian government ratified the 1997 KP on 12 December 2007; it entered into force on 11 March 2008.

[330] G. Bryner, 'The rapid evolution of climate change law,' *Utah B.J.* 22, 2007, p. 22; J. Gupta, 'Legal steps outside the Climate Convention: litigation as a tool to address climate change,' *RECIEL* 16(1) 2007, p. 78; J. Peel, 'The role of climate change litigation in Australia's response to global warming,' 24 *EPLJ* 90, 2007, p. 90; J.P. Pidot, 'Global warming in the courts. An Overview of current litigation and common legal issues' (online paper), Georgetown Environmental Law & Policy Institute, Georgetown University Law Center, 2006, p. 1, available from:
<www.law.georgetown.edu/gelpi/current_research/documents/GWL_Report.pdf> (last visited on 6 April 2009).

litigation is not only indicative of public demand for legal action[331] but can also be seen as a way 'to attract public attention and pressure [governments] to reach political solutions, including treaties and domestic laws.'[332] Peel has described climate change litigation as 'a promising strategy for raising the profile of the global warming issue and forcing business and governments to change their decision-making practices with regard to GHG emissions.'[333] Despite the fact that in many countries, adequate legal instruments for adjudicating global warming claims may be lacking, some authors anticipate that climate change litigation will keep spreading to other jurisdictions. [334] Whereas domestic litigation may eventually force the reluctant governments to act, it has been observed that it is unlikely to have any significant effect on the climate change problem. [335] Notwithstanding the fact that the overall general impact of litigation may be quite small, many climate change actions have been brought before national courts in an attempt to test various legal pathways to liability.

The present analysis has been limited to a selection of claims that broadly relate to liability, which is why many climate change-related actions, although significant in the global warming context, fall outside the scope of this chapter. For example, the human rights law-based *Shell Nigeria*[336] case, in which the applicants raised concerns about gas-flaring in Nigeria and alleged violations of the right to life (including the right to a healthy environment) and the right to human dignity, is not included in this study. Likewise excluded is the 2006 Inuit people's 'Petition to the Inter-American Commission on Human Rights Seeking Relief from Violations Resulting from Global Warming Caused by Acts and Omissions of the United States.'[337] Windmill cases are not addressed in the present analysis as plaintiffs' claims have not been based on climate change concerns.[338] Other examples of

---

[331] Hsu 2008, p. 718.

[332] E.A. Posner, 'Climate change and international human rights litigation: a critical appraisal,' 155 *U. Pa. L. Rev* 1925, 2007, p. 1944.

[333] Peel 2007, p. 103.

[334] M.G. Faure, A. Nollkaemper & Amsterdam International Law Clinic, *Climate Change Litigation Cases*, Amsterdam: Milieudefensie, 2007, p. 59.

[335] B. Preston, 'Climate change litigation,' paper delivered to Climate Change Governance after Copenhagen Conference, Hong Kong, 4 November 2010, p. 1.

[336] *Jonah Gbemre v. Shell Petroleum Development Company Nigeria Ltd.,* Suit No. FHC/B/CS/53/05, Federal High Court of Nigeria, Benin Judicial Division, 2005.

[337] 'Inuit Climate Change Petition Rejected,' *The New York Times*, available from: <www.nytimes.com/2006/12/16/world/americas/16briefs-inuitcomplaint.html> (last visited on 5 July 2012); full text of the petition available from: <www.inuitcircumpolar.com/files/uploads/icc-files/FINALPetition ICC.pdf> (last visited on 5 July 2012).

[338] *E.g. Genesis Power Ltd v. Franklin District Council* [2005] NZRMA 541; *Meridian Energy Limited & Ors v. Wellington City Council and Wellington Regional Council* [2007] W031/2007; *Unison Networks Ltd v. Hastings District Council* [2007] NZHC 1435.

climate change-related cases falling outside the scope of the present research include *Dimmock v. Secretary of State for Education and Skills*,[339] which involved an application to declare unlawful a decision by the Secretary of State for Education and Skills to distribute to every state secondary school in the United Kingdom (UK) a copy of Al Gore's film, 'An Inconvenient Truth;' and the German *BUND and Germanwatch* case, in which the applicants challenged the decision of the Minister for Economy and Labour denying the applicants' request for information on energy production projects.[340]

In order to identify some cross-cutting challenges shared by a multitude of climate change actions in different jurisdictions, the methodological approach adopted by this chapter is based on the type of relief sought by the plaintiffs. On that basis, three general categories of climate change liability lawsuits have been analysed: (1) claims related to procedural injury (including challenges to environmental impact assessments (EIAs) and judicial review cases); (2) claims for injunctive and/or declaratory relief; and (3) claims for compensation. This structure echoes the approach used in the analysis of interstate liability in Chapter 3, which is based on various forms of redress involving the obligation to pay compensation; the obligation to ensure prompt, adequate, and effective compensation; the obligation to negotiate a redress settlement; and the obligation to take response measures. It also resonates with the discussion of the content of state responsibility, *i.e.* the obligations of cessation, and reparation for injury in the form of restitution, compensation or satisfaction. The obligation of cessation and non-repetition is of particular significance as in practice it may be seen to correspond with injunctive relief in municipal legal systems (see Chapters 3 and 5). Within each category, the presentation of cases is chronological, unless stated otherwise.[341]

It has been the author's initial aim to cover all cases related to procedural injury, injunctive relief and actions for damages in all jurisdictions, on which information had been published in English before 1 October 2012. However, in the course of the research, the number of procedural injury cases proliferated.[342]

---

[339] *Dimmock v. Secretary of State for Education and Skills* [2007] EWHC 2288 (Admin).

[340] *Bund für Umwelt- und Naturschutz Deutschland e.V. (BUND) & Germanwatch e.V. gegen Bundesminister für Wirtschaft und Arbeit*, Verwaltungsgericht Berlin, Beschluss VG 10 A 215.04, 2006; unofficial English translation available from: <www.climatelaw.org/cases/case-documents/germany/de-export-jan06-eng.doc> (last visited on 9 October 2012).

[341] Where the decision of a lower court is appealed, the year of the decision on appeal is used to date the case for the purposes of the order of presentation.

[342] See *e.g. Habitat and Watershed Caretakers v. City of Santa Cruz*, 149 Cal.Rptr.3d 574 (Cal.App. 6 Dist. 2012); *Black Mesa Water Coalition v. Salazar*, 2012 WL 2848437 (D.Ariz. 2012); *Healdsburg Citizens for Sustainable Solutions v. City of Healdsburg*, 206 Cal.App.4th 988 (Cal.App. 1 Dist. 2012); *Coalition for a Sustainable Future in Yucaipa v. City of Yucaipa*, 198 Cal.App.4th 939 (Cal.App. 4 Dist. 2011); for more

Consideration of the facts and findings of all these cases would be repetitious and not conducive to a reader-friendly reflection of the research results. For that reason, a selection of procedural injury cases that represent different jurisdictions and different procedural injury issues has been made.

Pursuant to and relying upon the case law analysis, the legal problems associated with climate change liability are identified for each type of legal action. First, procedural injury claims are examined; next, the chapter considers claims for injunctive and/or declaratory relief; and finally, claims for compensation are dealt with, followed by some concluding remarks. Climate change lawsuits brought to date have posed profound challenges to plaintiffs. Legal obstacles to liability have included standing, causation, attribution of an act to the defendant, retroactivity, and other, mostly country-specific, procedural difficulties, such as non-justiciability of political questions.

4.2 Identification of Legal Challenges in Climate Change Claims

4.2.1 Claims Related to Procedural Injury

Claims related to procedural injury, including challenges to EIAs and judicial review cases, can be seen as the most successful category of climate change claims,[343] although procedural justice does not offer any immediate relief to plaintiffs as it is based on procedural, and not actual, injury. Procedural injury complaints do not directly challenge defendants' actions contributing to global warming, which helps plaintiffs to overcome standing problems and other obstacles inherent in claims for compensation and injunctions (see Sections 4.2.2 and 4.2.3). Legal actions in this category are often brought under administrative law as judicial review cases or cases challenging the competent authority's non-compliance with certain procedural requirements, such as EIA-related duties. Plaintiffs in such cases may also be challenging the content of certain EIAs insofar as global warming is not considered or is taken into account to an insufficient degree by the relevant agencies.

4.2.1.1 Political Question Doctrine

Rooted in the idea of the separation of powers, the political question doctrine is based on the notion that the judiciary must not intervene in policy issues

---

procedural injury cases, see also Columbia Center for Climate Change Law; <www.climatecasechart.com/> (last visited 11 January 2013).
[343] V. Tice, 'From Vermont's maples to Wybong's olives: cross-cultural lessons from climate change litigation in the United States and Australia,' 10 *Asian-Pac. L. & Pol'y J.* 292, 2008, p. 295.

appropriately left to be decided by the government.[344] Generally, procedural injury claims have not been hampered by the political question doctrine to the same degree as claims for compensation or injunctive relief.

In 1994, an Australian court was cautious in deciding the first global warming claim to be filed in Australia – *Greenpeace Australia v. Redbank Power Company*. The case concerned an appeal brought against the decision to grant development consent for the construction of a coal-fired power plant. Greenpeace complained that the proposed development would increase the total quantity of carbon dioxide emitted in New South Wales and thus contribute to the greenhouse effect. The applicant invited the court to apply the precautionary principle and refuse to grant consent to the proposed development. The court held that CO_2 emitted by the project would contribute to the greenhouse effect, which was 'a matter of national and international concern.'[345] In reference to the UNFCCC's work, the court noted that the response to the enhanced greenhouse effect was in the realm of governmental policy. It found that there was uncertainty as to what impact the project's emissions would have on global warming. The court admitted that the application of the precautionary principle dictated that a cautious approach be adopted and stressed that nonetheless it did not require that the greenhouse issue outweigh all other issues, which is why the development was approved.

The UNFCCC entry into force in 1994 and Australia's ratification of the KP in 2007 have clarified governmental policy on climate change. This, together with the fact that scientific certainty about the causes and effects of climate change has substantially increased since 1994, undoubtedly affected a 'greener' outcome of several cases subsequently decided in Australia.

4.2.1.2 Standing

Standing, or *locus standi*, is a threshold requirement that, if proven, enables the court to hear a case.[346] Standing 'addresses whether a party to a law suit is a proper party to sue, and does not address whether the asserted claim is appropriate;' it is 'one factor in determining whether a suit is legitimately justiciable in court.'[347] If a plaintiff is found to have no standing, their case will not reach the merits stage of the proceedings. Standing in climate change cases has often amounted to a serious

---

[344] For a detailed discussion of the political question doctrine, see Sections 4.2.2.1 and 4.2.3.1.

[345] *Greenpeace Australia Ltd v. Redbank Power Company Pty Ltd* [1994] NSWLEC 178, conclusion.

[346] F. Benzoni, 'Environmental standing: who determines the value of other life?', 18 *Duke Envtl. L. & Pol'y F.* 347, 2008, p. 347.

[347] B.C. Mank, 'Standing and future generations: does Massachusetts v. EPA open standing for generations to come?', 34 *Colum. J. Envtl. L.* 1, 2009, p. 24.

hurdle, particularly in US courts; however, claims related to procedural injury have not been hindered to the same degree as claims for injunctive or compensatory relief.

Article III of the US Constitution, limiting the power of the courts to hearing cases as opposed to giving advisory opinions, does not contain specific requirements for standing; the so-called Article III standing requirements developed in case law. Referring to its earlier decision in *Lujan v. Defenders of Wildlife*,[348] the US Supreme Court formulated them as follows:

> [T]o satisfy Article III's standing requirements, a plaintiff must show (1) it has suffered an 'injury in fact' that is (a) concrete and particularized and (b) actual or imminent, not conjectural or hypothetical; (2) the injury is fairly traceable to the challenged action of the defendant; and (3) it is likely, as opposed to merely speculative, that the injury will be redressed by a favorable decision.[349]

In establishing these requirements, the onus is on the plaintiff.

As mentioned earlier, standing has not been a major obstacle for plaintiffs in procedural injury claims. For example, in *Friends of the Earth v. Mosbacher,* the court explicitly stated that the standing requirements were less stringent in procedural injury cases. This case was initially known as *Friends of the Earth v. Watson*[350] when non-governmental organizations (NGOs) Friends of the Earth and Greenpeace and the cities of Boulder, Colorado; Arcata, California; and Oakland, California, filed a suit against two government agencies: the Overseas Private Investment Corporation (OPIC) and the Export-Import Bank of the US (Ex-Im). The plaintiffs alleged that the defendants provided financial assistance to projects that contributed to climate change without conducting EIAs required under the National Environmental Policy Act 1969 (NEPA). In their brief, the plaintiffs pointed out that the amount of carbon dioxide attributable to the projects supported by OPIC and Ex-Im was 'much higher than the entire amount of CO_2 that was released from the *worldwide* consumption of petroleum, natural gas, coal, and the flaring of natural gas in the year 2000.'[351] The plaintiffs challenged OPIC's and Ex-Im's determinations that the projects they supported did not have a significant environmental impact. The court found that the plaintiffs had standing to bring their

---

[348] *Lujan v. Defenders of Wildlife*, 112 S.Ct. 2130 (U.S.Minn. 1992).

[349] *Friends of the Earth, Inc. v. Laidlaw Environmental Services, Inc.*, 120 S.Ct. 693 (U.S. 2000), p. 704, footnote omitted.

[350] *Friends of the Earth, Inc. v. Watson*, 35 ELR 20179 (N.D.Cal. 2005).

[351] 2004 WL 5584704 (Trial Pleading) Complaint for Declaratory and Injunctive Relief (Second Amended) Administrative Procedure Act Case (Jan. 16, 2004), *Friends of the Earth, Inc. v. Watson*, para. 151, emphasis in the original.

claim noting the lower threshold for causation in procedural injury cases. It found that the plaintiffs demonstrated an injury in fact by presenting the evidence that the emissions from the projects supported by the defendants would threaten their concrete interests. Given the lower causation and redressability standards in procedural injury cases, the court found the plaintiffs to have sufficiently alleged both. It is significant that under the procedural-right analysis, the redressability and immediacy requirements are relaxed.[352] In other words, in order to establish procedural standing, a plaintiff only has to show that they have a concrete procedural interest that is threatened. As the Supreme Court noted in *Lujan v. Defenders of Wildlife*, standing as a procedural right has to satisfy the following elements: (1) 'the plaintiff must be a person 'accorded a procedural right to protect his concrete interest'' and (2) 'the plaintiff must have some concrete interest threatened that is the ultimate foundation of his or her standing.'[353]

Two years later, the court assessed the merits of the case ruling that the plaintiffs were not entitled to an injunction directing the defendants to make an Environmental Impact Statement (EIS) for every fossil fuel project they might approve in the future.[354] Under NEPA, only 'major Federal actions significantly affecting the quality of the human environment' necessitate an EIS. Therefore, the court found it impossible to conclude as a matter of law that 'every fossil fuel related project that Ex-Im or OPIC may undertake in the future will trigger NEPA's requirements.'[355] Eventually, the case was settled in February 2009. Under the settlement, Ex-Im will begin to take carbon dioxide emissions into consideration when deciding whether or not to approve transactions related to fossil fuel projects. Ex-Im will also develop and implement a carbon policy and promote consideration of climate change issues.[356] According to the settlement, OPIC will adopt a policy of reducing by 20 per cent over the next 10 years GHG emissions associated with projects that emit more than 100 thousand tons of CO_2 equivalents per year.[357]

In another US case, *Center for Biological Diversity v. Interior*,[358] the petitioners advanced two theories of standing: substantive and procedural. The case concerned a Leasing Program of the US Department of Interior directed at the expansion of

---

[352] M. Sugar, 'Massachusetts v. Environmental Protection Agency,' 31 *Harv. Envtl. L. Rev.* 531, 2007, p. 538.

[353] R.S. Abate, 'Massachusetts v. EPA and the future of environmental standing in climate change litigation and beyond,' 33 *Wm. & Mary Envtl. L. & Pol'y Rev.* 121, 2008, p. 126, citing *Lujan v. Defenders of Wildlife*, 112 S.Ct. 2130 (U.S.Minn. 1992).

[354] *Friends of the Earth, Inc. v. Mosbacher*, 488 F.Supp. 2d 889 (N.D.Cal. 2007).

[355] *Friends of the Earth, Inc. v. Mosbacher*, 488 F.Supp. 2d 889 (N.D.Cal. 2007), p. 919.

[356] <www.climatelaw.org/cases/case-documents/us/exim.pdf> (last visited on 3 June 2009).

[357] <www.foe.org/pdf/OPIC_Settlement.pdf> (last visited on 3 June 2009).

[358] *Center for Biological Diversity v. United States Department of the Interior*, 563 F. 3d 466 (D.C.Circ. 2009).

leasing areas within the Outer Continental Shelf. The petitioners challenged the approval of this Leasing Program by the Secretary of the Interior. Among other things, the petitioners argued that the Leasing Program violated both the Outer Continental Shelf Lands Act (OCSLA) and the National Environmental Policy Act (NEPA) because Interior had failed to consider the effects of climate change.[359]

Distinguishing the case from *Massachusetts v. EPA* (see Section 4.2.2.2 below), the court found that the petitioners lacked standing on their substantive theory failing to establish the injury and causation. However, the court found that the petitioners did have standing on their procedural theory because they had shown that they possessed 'a threatened particularized interest, namely their enjoyment of the indigenous animals of the Alaskan areas listed in the Leasing Program.'[360]

On the merits, the court found that the petitioners' NEPA challenges were not ripe for review because the plaintiffs' rights had not been implicated since no lease-sales had yet occurred at the moment they filed their petitions. As regards the plaintiffs' justiciable OCSLA-based claims, they were found to lack merit and as such had to fail. However, the court concluded that the petitioners' challenge to the Leasing Program on grounds that the Program's environmental sensitivity rankings were irrational was meritorious. Thus, the Leasing Program was vacated and remanded to the Secretary of the Interior for reconsideration.

In Australian environmental law, standing depends on the nature of the claim, the relief sought, and any applicable law as well as the court's assessment of the circumstances of the case. Some statutes, *e.g.* the Heritage Properties Conservation Act 1983, establish very broad standing requirements: 'any interested person' is entitled to bring a claim. Common law principles regarding standing apply in States and Territories that do not have administrative law procedure statutes, and the test of whether a plaintiff has a 'special interest in the subject matter' is used.[361] The requirements for having a 'special interest' have been expressed in *Australian Conservation Foundation v. Commonwealth.* A person has a 'special interest' if:

(1) the person or group will suffer actual or apprehended injury to some material or non-material spiritual or cultural interest(s);

(2) the person or group has been specially affected to a substantially greater degree or a significantly different manner than the general public; and

---

[359] *Center for Biological Diversity v. United States Department of the Interior*, 563 F. 3d 466 (D.C.Circ. 2009), p. 471.

[360] *Center for Biological Diversity v. United States Department of the Interior*, 563 F. 3d 466 (D.C.Circ. 2009), p. 479.

[361] J. Smith & D. Shearman, *Climate Change Litigation: Analysing the Law, Scientific Evidence & Impacts on the Environment, Health & Property*, Adelaide: Presidian Legal Publications, 2006, p. 58.

(3) there exists a relationship of a sufficient proximity between the interest and the plaintiff individual or group.[362]

Standing has not been at issue in Australian climate change cases because most of Australia's global warming legal actions have been brought as requests for judicial review under particular statutes and the applicants' standing has not been challenged.

Thus, in procedural injury cases analysed above, standing cannot be said to inhibit climate change plaintiffs as allegations of procedural injury are aided by a standing threshold lower than when actual injury is alleged.

4.2.1.3 Causation

Causation in climate change cases has been described as 'the greatest obstacle to the majority of plaintiffs.'[363] In the US climate change litigation, the problem of causation has generally been two-fold. First, causation must be proven as an element of standing: the injury has to be 'fairly traceable' to the actions of a defendant (see Section 4.2.1.1 above). Second, causation would have to be proven at the merits stage of the proceedings. It is apparent that the standard of proof in the latter case must be higher than that for establishing standing; yet, few climate change liability cases have been argued on the merits. Causation requirements for demonstrating standing in procedural injury cases are lower than in cases alleging actual injury and 'environmental plaintiffs need not show that substantive environmental harm is imminent.'[364]

In *Natural Resources Defense Council v. Kempthorne*, the plaintiffs challenged a 2005 Biological Opinion issued by the US Fish and Wildlife Service pursuant to the Endangered Species Act. In this judicial review case, the plaintiffs did not need to demonstrate Article III standing. This case concerned the effect on a threatened species of fish, the Delta smelt, of the coordinated operation of the federally managed Central Valley Project (CVP) and California's State Water Project (SWP). CVP and SWP are among the world's largest water diversion projects.[365] The plaintiffs moved for summary judgment on the grounds, *inter alia*, that the

---

[362] Smith & Shearman 2006, p. 58 citing *Australian Conservation Foundation v. Commonwealth* [1980] 146 CLR 493.

[363] Smith & Shearman 2006, p. 107. Due to differences in approaches to causation between different jurisdictions, claims in the current section are presented in the following way: chronologically organized US claims come first and claims from Australia and New Zealand come second, also organized in the chronological order.

[364] *Friends of the Earth, Inc. v. Watson*, 35 ELR 20179 (N.D.Cal. 2005).

[365] *Natural Resources Defense Council v. Kempthorne*, 506 F.Supp. 2d 322 (E.D.Cal. 2007), p. 328.

biological opinion failed to consider the possible effects of climate change on the smelt's habitat.[366] The plaintiffs argued that despite the evidence that climate change could seriously impact the smelt, the biological opinion had not taken into account its probable effects. The court noted that while the precise magnitude of climate change-related changes was uncertain, 'judgments about the likely range of impacts [could] and [had] been made.'[367] The court expressed some doubt as to the precise impacts of climate change, ruling that the likely range of such impacts had nonetheless to be taken into account. The court found in favour of the plaintiffs as to their climate change claim stating that the 'absence of *any* discussion in the [biological opinion] of how to deal with any climate change is a failure to analyse a potentially "important aspect of the problem."'[368]

In *Center for Biological Diversity v. Interior*, causation had to be established as an element of standing. In this case, the plaintiffs were afforded procedural standing whereby the causation and redressability elements of the Article III standing test were relaxed. Meanwhile, in rejecting the plaintiffs' substantive theory of standing, the court observed:

> In order to reach the conclusion that Petitioners are injured because of Interior's alleged failure to consider the effects of climate change with respect to the Leasing Program, Petitioners must argue that: adoption of the Leasing Program will bring about drilling; drilling, in turn, will bring about more oil; this oil will be consumed; the consumption of this oil will result in additional carbon dioxide being dispersed into the air; this carbon dioxide will consequently cause climate change; this climate change will adversely affect the animals and their habitat; therefore Petitioners are injured by the adverse effects on the animals they enjoy. *Such a causal chain cannot adequately establish causation because Petitioners rely on the speculation that various different groups of actors not present in this case – namely, oil companies, individuals using oil in their cars, cars actually dispersing carbon dioxide – might act in a certain way in the future.*[369]

Similarly, due to a lower standard of proof, the plaintiffs in *Friends of the Earth v. Watson*[370] were able to establish procedural standing having shown that it was 'reasonably probable' that the challenged action would threaten their concrete interests.[371]

---

[366] *Natural Resources Defense Council v. Kempthorne*, 506 F.Supp. 2d 322 (E.D.Cal. 2007), p. 348.

[367] *Natural Resources Defense Council v. Kempthorne*, 506 F.Supp. 2d 322 (E.D.Cal. 2007), p. 368.

[368] *Natural Resources Defense Council v. Kempthorne*, 506 F.Supp. 2d 322 (E.D.Cal. 2007), p. 370, emphasis in the original.

[369] *Center for Biological Diversity v. United States Department of the Interior*, 563 F. 3d 466 (D.C.Circ. 2009), pp. 478-479, emphasis added.

[370] *Friends of the Earth, Inc. v. Mosbacher*, 488 F.Supp. 2d 889 (N.D.Cal. 2007).

[371] *Friends of the Earth, Inc. v. Watson*, 35 ELR 20179 (N.D.Cal. 2005).

Australian courts have traditionally applied 'the common sense and experience test encompassing the 'but for' test,' *i.e.* the court would consider whether or not the plaintiff's damage would have occurred but for the defendant's actions. The plaintiff is required to show that the defendant's acts caused the damage suffered and there must be 'a more probable inference in favour of what is alleged' and not just a mere possibility.[372] As regards climate change, it is 'difficult to identify, on the balance of probabilities, that the greenhouse gas emissions of the defendant caused the harm suffered in the presence of [...] scientific doubt.'[373]

For the purposes of clarifying climate change causation in claims related to procedural injury as dealt with in Australian and New Zealand courts, it is also helpful to distinguish between general causation and specific causation. The former 'requires proof that anthropogenic emissions cause changes in radiative forcing and thus the global climate,' while the latter 'requires proof that a particular impact or injury is attributable to (particular) anthropogenic emissions or to the global warming caused by them.'[374] It is essential to separate (a) the causal link between overall anthropogenic emissions and climate change from (b) climate change damage being caused by emissions from a specific source, if only by way of contribution. General causation can thus be regarded as a prerequisite for specific causation. Albeit application of the precautionary principle by Australian courts, as will be demonstrated below, has in some instances resulted in a lower standard of proof, showing specific causation has posed considerable challenges in a number of Australian procedural injury cases.

In light of new evidence published in the 2007 IPCC AR4, scientific doubt can no longer be used as a ground for rejecting the general causal link between GHG emissions, climate change, and its injurious manifestations (see Chapter 1). Correspondingly, as this section will show, the Australian and New Zealand courts' view of causation in climate change claims related to procedural injury has gradually evolved.[375]

The Environment Court of New Zealand was liberal in its approach in *Environmental Defence Society v. Auckland Regional Council* and accepted both the general causal connection between GHG emissions and climate change, and the specific link between climate change and emissions from a particular source. This

---

[372] N. Durrant, 'Tortious liability for greenhouse gas emissions? Climate change, causation and public policy considerations,' *QUT LJJ* 23, 2007.

[373] Durrant 2007.

[374] R. Verheyen, *Climate Change Damage and International Law – Prevention Duties and State Responsibility*, Leiden: Martinus Nijhoff Publishers, 2005, p. 257.

[375] Cf. *Greenpeace Australia Ltd v. Redbank Power Company Pty Ltd* [1994] NSWLEC 178 with, *e.g. Gray v. Minister for Planning and Ors* [2006] NSWLEC 720 or *Gippsland Coastal Board v. South Gippsland SC & Ors (No 2) (includes Summary) (Red Dot)* [2008] VCAT 1545 where the court applied the precautionary principle.

appeal involved an application for resource consent to construct and operate a 400 megawatt gas fired combined cycle power station at Otahuhu C in South Auckland. Since the air discharge consent granted by the Regional Council included no condition addressing the discharge of any GHGs, the Environmental Defence Society sought an imposition of such a condition 'to offset all the carbon dioxide emissions by the planting of trees to act as "carbon-sinks."'[376] The Environment Court analysed New Zealand's international commitments under the UNFCCC and the Kyoto Protocol and concluded that New Zealand was under an obligation to reduce its GHG emissions by, *inter alia*, using carbon sinks. On the evidence presented, the court found that

> the greenhouse effect and the possibility of climate change are a matter of serious concern. It is difficult to assess the degree of concern because there are widely differing opinions as to the likely environmental consequences. However the weight of scientific opinion is such, that on balance, the threat posed by the enhanced greenhouse effect is sufficiently significant for us to conclude that the green house effect is likely to result in significant changes to the global environment, including New Zealand and the Auckland region.[377]

The Environment Court dismissed the appeal and did not impose on the air discharge consent the condition requested by the Environmental Defence Society. The court stressed its inability to assess the reasonableness and appropriateness of the proposed condition. However, the court accepted that 'the present scientific consensus [was] that the cumulative anthropogenic emissions of carbon dioxide on a global basis contribute[d] to climate change.'[378] It also noted that while it was not possible to definitively quantify, the prognosis was sufficiently serious for the court to find that 'the proposed emissions from Otahuhu C [would] result, in a cumulative way, in an adverse effect of some consequence.'[379] In accepting the scientific consensus with regard to anthropogenic emissions contributing to climate change and finding that emissions from a given source would cumulatively result in an adverse effect, the court recognized the specific causal link.

The causal link between future emissions from a particular source and the resultant adverse effects upon the world climate was stressed in *Australian Conservation Foundation v. Minister for Planning*. In this judicial review application, the Victorian Civil and Administrative Tribunal (VCAT) considered the question whether a planning panel could refuse to consider the environmental effects of GHG emissions resulting from continued operation of the Hazelwood

---

[376] *Environmental Defence Society (Inc) v. Auckland Regional Council* [2002] NZRMA 492, para. 4.
[377] *Environmental Defence Society (Inc) v. Auckland Regional Council* [2002] NZRMA 492, para. 65.
[378] *Environmental Defence Society (Inc) v. Auckland Regional Council* [2002] NZRMA 492, para. 88.
[379] *Environmental Defence Society (Inc) v. Auckland Regional Council* [2002] NZRMA 492, para. 88.

Power Station. Brown coal reserves resourcing the Hazelwood Power Station were due to run out in 2009 and the owner applied to extend the power station's operation until 2031. A panel set up to consider the proposal was instructed by the Minister not to consider GHG emissions from the power station. Environmental groups who had made submissions on the issue of GHG emissions brought proceedings to the VCAT challenging the exclusion from the panel's consideration of GHG-related matters. The court found that the environmental impacts of GHG emissions were relevant and that the panel should have considered 'submissions to the effect that the continuation of the Hazelwood Power Station may have adverse environmental effects by reason of the generation of greenhouse gases' as required under the Planning and Environment Act, notwithstanding the Minister's instructions.[380]

In considering the approval of the Hazelwood Power Station's continued operation, the court found that there was

> [...] a sufficient nexus between the approval of Amendment C32 and the environmental effect of greenhouse gases that are likely to be produced by the use of the Hazelwood Power Station beyond 2009. [...] Put another way, the approval of Amendment C32 will make it more probable that the Hazelwood Power Station will continue to operate beyond 2009; which, in turn, may make it more likely that the atmosphere will receive greater greenhouse gas emissions than would otherwise be the case; which may be an environmental effect of significance.[381]

Thus, the court recognized the role of a single GHG-intensive project in causing a significant environmental effect by way of emissions released in the course of its operation.

The two types of causation have been satisfactorily established in *Gray v. Minister for Planning,*[382] also known as the Anvil Hill case.[383] The applicant challenged an EIA carried out for the construction of a coal mine on Anvil Hill alleging that (1) the assessment did not comply with the environmental assessment requirements and that (2) the Director-General of the Department of Planning failed to take into account the principles of ecologically sustainable development (ESD), *i.e.* the precautionary principle and intergenerational equity. The court noted that the issue of causation arose with regard to the second point of challenge. In finding

---

[380] *Australian Conservation Foundation v. Minister for Planning* [2004] VCAT 2029, para. 49; the Minister's issuance of terms of reference to the panel directing it not to consider GHG-related matters was addressed through a separate process.

[381] *Australian Conservation Foundation v. Minister for Planning* [2004] VCAT 2029, paras 46-47.

[382] *Gray v. Minister for Planning and Ors* [2006] NSWLEC 720.

[383] *Gray v. Minister for Planning and Ors* [2006] NSWLEC 720, for a detailed analysis see A. Rose, 'Gray v Minister for Planning: the rising tide of climate change litigation in Australia,' *Syd. L. Rev.* 28, 2007.

that there was 'a sufficiently proximate link between the mining of a very substantial reserve of thermal coal [in New South Wales and] the emission of GHG which contribute[d] to climate change/global warming,'[384] the court observed:

> Climate change/global warming is widely recognised as a significant environmental impact to which there are many contributors worldwide but the extent of the change is not yet certain and is a matter of dispute. The fact there are many contributors globally does not mean that the contribution from a single large source such as the Anvil Hill Project in the context of NSW should be ignored in the environmental assessment process. The coal intended to be mined is clearly a potential major single contributor to GHG emissions deriving from NSW given the large size of the proposed mine. That the impact from burning the coal will be experienced globally as well as in NSW, but in a way that is currently not able to be accurately measured, *does not suggest that the link to causation of an environmental impact is insufficient.*[385]

Although the assessment was found to have been in conformity with the environmental assessment requirements, the court ruled that the applicant was successful in relation to the second argument because the Director-General had failed to take the precautionary principle into account. Recognizing the anthropogenic character of climate change (general causation), the court found that the Minister had failed to take the precautionary principle into account when carrying out an EIA for a GHG emitting mine. In view of the fact that one emitter was likely to be responsible only for a proportion of CO_2 contributing to climate change, the court stressed the importance of not ignoring a single source of GHG emissions because of its relatively small contribution to the global concentrations of GHGs in the atmosphere (specific causation).

The general causal link as well as specific causation appear to have been implicitly recognized in *Greenpeace v. Northland Regional Council & Mighty River Power*, an appeal launched against an earlier decision of the New Zealand Environment Court. Mighty River Power applied to the Northland Regional Council for resource consent to commission its facilities at Marsden B, a coal-fired power station, and Greenpeace opposed the application. The application was granted in large part by amending and striking out parts of Greenpeace's appeal.[386] The appeal in question was brought against that decision and concerned the correct interpretation of a legislative provision relating to discharge of GHGs, which prohibited authorities from considering the effect of GHG emissions on climate change in discharge and coastal permit applications 'except to the extent that the

---

[384] *Gray v. Minister for Planning and Ors* [2006] NSWLEC 720, para. 100.

[385] *Gray v. Minister for Planning and Ors* [2006] NSWLEC 720, para. 98, emphasis added.

[386] *Greenpeace v. Northland Regional Council & Mighty River Power Ltd* [2006] NZHC 1212, para. 6.

use and development of renewable energy enables a reduction in the discharge into air of greenhouse gases.' The High Court of New Zealand ruled in favour of the appellant and held that climate change was a relevant consideration irrespective of whether a permit application concerned renewable or non-renewable energy production.[387]

Yet, causation in climate change claims related to procedural injury has not uniformly been the subject of progressive interpretation by the judiciary. For example, in *Wildlife Whitsunday*, an Australian court was sceptical on the specific causation issue. This application for judicial review concerned challenges to the EIAs conducted for proposals for two new coal mines: the Isaac Plains and the Sonoma projects.[388] The applicant submitted that climate change was a matter of national environmental significance and that consideration had to be given to the impact on global warming from the coal to be extracted from the two mines. It was alleged by the applicant that the GHGs emitted as a result of the combustion of the coal from the two mines would contribute to global warming that, in turn, would negatively affect the Great Barrier Reef Heritage Area and the Shoalwater and Corio Bays Ramsar site. The applicant invoked the Convention on Biological Diversity, the World Heritage Convention, the Ramsar Convention, and the UNFCCC and, relying heavily on the latter agreement, stressed Australia's obligations under those treaties. The applicant's argument was that the respondent failed to take into account the adverse impacts the two projects were likely to have due to emissions from the mined coal contributing to global warming. The court found that, in approving the applications, the Minister had in fact taken the environmental effects of GHG emissions into account. None of the numerous grounds for review presented by the applicant were established. With regard to specific causation, the court noted that it was 'far from satisfied that the burning of coal at some unidentified place in the world, the production of greenhouse gases from such combustion, its contribution towards global warming and the impact of global warming upon a protected matter, [could] be so described.'[389]

Specific causation remained problematic in *Re Xstrata Coal Queensland*.[390] Xstrata, a mining company, applied to the tribunal for a license to extend the surface area of an open cut coal mine and for the grant of the related environmental authority, *i.e.* a

---

[387] *Greenpeace v. Northland Regional Council & Mighty River Power Ltd* [2006] NZHC 1212, para. 57.
[388] *Wildlife Preservation Society of Queensland Proserpine/Whitsunday Branch Inc. v. Minister for the Environment & Heritage & Ors* [2006] FCA 736.
[389] *Wildlife Preservation Society of Queensland Proserpine/Whitsunday Branch Inc. v. Minister for the Environment & Heritage & Ors* [2006] FCA 736, para. 72.
[390] *Re Xstrata Coal Queensland Pty Ltd & Ors* [2007] QLRT 33.

mining lease. The tribunal was to make a recommendation to the Minister on the application. Objections were filed by the Queensland Conservation Council that an increase in the mining activity would lead to an increase in GHG emissions that contributed to global warming and climate change.[391] Queensland Conservation Council requested that conditions be imposed on the grant of the mining licence to 'avoid, reduce or offset the emissions of greenhouse gases that are likely to result from the mining, transport and use of the coal from the mine.'[392] The tribunal's presiding member recommended to the Minister that the application be granted as a whole, without any conditions sought by the objectors as he saw no specific causal link between Xstrata's GHG emissions and harm caused by global warming and climate change.[393] Notably, the Tribunal did not afford much weight to the IPCC Fourth Assessment Report and, particularly, the Stern Review on the Economics of Climate Change dismissing it as 'biased, selective and unbalanced' and 'scientifically flawed,'[394] which is indicative of a measure of scepticism also with regard to general causation.

On appeal, although it was not the function of the court to validate the tribunal's analysis and methodology, the Queensland Court of Appeal nevertheless stated:

> The fact that climate change is occurring and that anthropogenic greenhouse gas emissions have contributed to it, was undoubtedly common ground between the parties at the hearing. [...] What was in issue was the extent to which the proposed mine would contribute to global warming and whether, in the applicable factual and statutory matrix, the Tribunal should impose conditions on the recommended granting of Xstrata's applications in response to the mine's potential contribution to global warming.[395]

This inference is significant in that it reaffirms that the general causal link between GHG emissions and climate change was not challenged.

Appreciation of the threat posed by the rising sea levels caused by climate change was at the crux of the court's decision in *Charles & Howard v. Redland Shire Council*, whereby the general causal link between climate change and the rising sea levels and the impacts of the latter on the specific proposed development were recognized. The Supreme Court of Queensland upheld a lower court's

---

[391] *Re Xstrata Coal Queensland Pty Ltd & Ors* [2007] QLRT 33, objections.

[392] *Re Xstrata Coal Queensland Pty Ltd & Ors* [2007] QLRT 33, para. 8.

[393] *Re Xstrata Coal Queensland Pty Ltd & Ors* [2007] QLRT 33, para. 21; recommendation.

[394] *Re Xstrata Coal Queensland Pty Ltd & Ors* [2007] QLRT 33, para. 16 referring to R. Carter's and I. Byatt's critique of the Stern Review in: *World Economics*, Vol. 7, No. 4, October – December 2006, pp. 165-232.

[395] *Queensland Conservation Council Inc. v. Xstrata Coal Queensland P/L & Ors* [2007] QCA 338, para. 41.

decision that the impact of climate change on sea levels in a flood susceptible area proposed to be filled for residential development justified a condition of relocating the proposed development to an area less prone to tidal inundation. The Council approved the applicant's development application but imposed several conditions, one of which prohibited siting the proposed development at the applicant's location of preference citing climate change-related tidal inundation. One of the questions put to the appellate judges was whether the primary judge had erred in considering evidence of climate change when denying the applicant's appeal. The Court found that

> His Honour was entitled, as he did, to take into account [...] the impact of climate change on sea levels on the area proposed to be filled by the applicant and on the area proposed by the Council in its disputed condition, and to accept [...] that the applicant's proposed building site may be vulnerable to rising sea levels because of climate change.[396]

The general causal link between climate change and human GHG emissions was also recognized by the court in *Walker v. Minister for Planning*. In these judicial review proceedings, the applicant challenged the Minister's approval of a concept plan for a residential development at Sandon Point – a site of 'significant inherent cultural, ecological and social' value.[397] Importantly, one of the grounds upon which the applicant challenged the Minister's decision was that the Minister failed to take into consideration 'the principles of ecologically sustainable development (ESD) and the impact of the proposal upon the environment in several respects, including whether the flooding impacts of the project would be compounded by climate change.'[398] After giving careful consideration to the development of the concept of ESD globally and in Australia from the 1972 United Nations Conference on the Human Environment in Stockholm onwards, the court proceeded to review a wide range of ESD cases and addressed ESD in relation to climate change specifically. It noted the IPCC's four climate change Assessment Reports focusing on Australia's potential vulnerability to 'climate change caused by global warming, including increasing coastal vulnerability to storm surges and sea level rises' addressed in the IPCC Third Assessment Report in 2001.[399] The court noted that scientific support for 'a link between a rise in global temperatures and an increase in the atmosphere in the concentration of greenhouse gases resulting from human activities' had been recognized by the courts of Australia, the US, and the UK.[400] The court extensively analysed climate change litigation in Australia and abroad

---

[396] *Charles & Howard Pty Ltd v. Redland Shire Council* [2007] QCA 200, para. 28.
[397] *Walker v. Minister for Planning* [2007] NSWLEC 741, para. 17.
[398] *Walker v. Minister for Planning* [2007] NSWLEC 741, para. 2 (emphasis omitted).
[399] *Walker v. Minister for Planning* [2007] NSWLEC 741, para. 125.
[400] *Walker v. Minister for Planning* [2007] NSWLEC 741, para. 126.

and noted a number of academic articles focusing on climate change litigation and legislation. It described climate change as a 'deadly serious issue' as evidenced by 'global scientific support' for its existence and the risks involved.[401] The court held that the Minister was under an implied obligation to consider the climate change flood risk when deciding whether to approve the concept plan. The case was subsequently appealed by the Minister, and the Supreme Court of New South Wales set aside the decision of the lower court. It held that although the Minister was required to have regard to the public interest in accordance with the Environmental Planning and Assessment Act, not having regard to the ESD principles did not necessarily breach this obligation.

> [...] [A] 'mandatory' requirement that the Minister have regard to the public interest is [not] necessarily breached in all cases where the Minister does not have regard to the principles of ESD. The 'mandatory' requirement that the Minister have regard to the public interest does not of itself make it mandatory (that is, a condition of validity) that the Minister have regard to any particular aspect of the public interest, such as one or more of the principles of ESD. Whether or not it is mandatory to have regard to one or more of the principles of ESD must depend on statutory construction.[402]

Yet, *Walker v. Minister for Planning* remains exemplary in terms of the lower courts' causation analysis as its reversal on appeal was largely rooted in statutory interpretation.

In the *Gippsland Coastal Board v. South Gippsland Red-Dot* case,[403] the court did not only recognize the general link between climate change and extreme weather events but went as far as to apply the precautionary principle in deciding that the sea level rise and storm surge would create a risk for the specific proposed development. The claim involved a set of applications for review concerning the decision of the council to grant a permit for a dwelling on land allotments located in a low-lying coastal area. The VCAT found that the land was unsuitable for residential development as it was at risk of inundation due to sea level rises resulting from climate change.[404] The tribunal accepted that there was a general consensus that 'some level of climate change [would] result in extreme weather conditions.' The VCAT found that 'sea level rise and risk of coastal inundation

---

[401] *Walker v. Minister for Planning* [2007] NSWLEC 741, para. 161.

[402] *Minister for Planning v. Walker* [2008] NSWCA 224, para. 44.

[403] The practice of the Victorian Civil and Administrative Tribunal (VCAT) is to designate cases of interest as 'Red Dot Decisions'. A summary is published and the reasons why the decision is of interest or significance are identified.

[404] *Gippsland Coastal Board v. South Gippsland SC & Ors (No 2) (includes Summary) (Red Dot)* [2008] VCAT 1545, introduction.

[were] relevant matters to consider in appropriate circumstances.'[405] Most significantly, the tribunal noted:

> 47. The relevance of climate change to the planning decision making process is still in an evolutionary phase. Each case concerning the possible effects of climate change will turn on its own facts and circumstances.
> 48. In the present case, we have applied the precautionary principle. We consider that increases in the severity of storm events coupled with rising sea levels create a reasonably foreseeable risk of inundation of the subject land and the proposed dwellings, which is unacceptable. [...]

Thus, the decisions of the responsible authority were set aside and no permits were granted.

4.2.1.4 Attribution

Attribution of a given act to a particular defendant is an issue closely related to causation, which may pose great difficulties to climate change plaintiffs. Generally, domestic courts have not extensively addressed the question of attribution to a defendant of (a proportion of) the harm caused by anthropogenic climate change. However, attribution is relevant to determining the proportion of the harm defendants should be held responsible for and whether and to what extent they should bear the costs. At present, it is not clear how national courts will go about apportioning liability between multiple contributing sources given the historic and geographical dimension of climate change, and in any case those determinations are not to be made in procedural decisions.

It has been emphasized that climate change is an inherently global phenomenon with multiple contributors across space and time. First, anthropogenic emissions have been on the rise since the beginning of the Industrial Revolution in the 1750s. Thus, the current levels of GHG concentration in the atmosphere are a cumulative result of historic emissions.[406] Second, many contributors to global warming have ceased to exist. The third important temporal characteristic of climate change is that not only is there a delay between GHG emissions release and a rise in temperatures. There is also a delay between a rise in temperatures and the impact associated with such a rise, as well as between the impact of the rising temperatures and climate change-related damage. Difficulties with attributing climate change-related damage to a particular emitter additionally complicate proof of causation.

---

[405] *Gippsland Coastal Board v. South Gippsland SC & Ors (No 2) (includes Summary) (Red Dot)* [2008] VCAT 1545, para. 46.
[406] IPCC *AR4, Synthesis Report* (2007), p. 37.

As regards the spatial dimension of climate change, it is obvious that every human being contributes to it by driving cars, heating houses or purchasing certain products and yet, not everyone is likely to be sued for it. Admittedly, some entities emit a great deal more GHGs than others but how does one attribute the harm caused by climate change to a given emitter? The fact that emitters are located the world over aggravates the problem. In *Environmental Defence Society v. Auckland Regional Council*, a New Zealand court approached attribution by focusing on the cumulative effect of GHG emissions from a given source. The court found that the emissions from a power station would 'result, in a cumulative way, in an adverse effect of some consequence.'[407] Whether the contribution to global climate change attributable to a given emitter is large or small, its effect cannot be disregarded.

From a procedural point of view, an Australian court resolved the issue of attribution by stating that at least as far as EIAs are concerned, the 'fact that there are many contributors globally does not mean the contribution from a single large source [...] should be ignored in the environmental assessment process.'[408]

4.2.2 Actions for Injunctive and/or Declaratory Relief

Actions seeking injunctions constitute another category of climate change cases. Injunctive relief – a court order requiring a party to do, or to refrain from doing, certain acts – is a form of relief ultimately related to climate change mitigation due to its preventive character. Lawsuits involving requests for injunctions are of great importance in terms of identifying hurdles in a potential international claim focusing on prevention. Later on, in Chapter 5, parallels are drawn between actions for injunctive and/or declaratory relief launched domestically and state responsibility claims seeking cessation and declaratory relief launched at the interstate level.

4.2.2.1 Political Question Doctrine

Actions for injunctive relief have been fraught with many more obstacles than claims related to procedural injury. For instance, the doctrine of non-justiciability of political questions has presented a considerable jurisdictional barrier to climate change plaintiffs. The US Supreme Court has indicated that 'disputes involving

---

[407] *Environmental Defence Society (Inc) v. Auckland Regional Council* [2002] NZRMA 492, para. 88.
[408] *Gray v. Minister for Planning and Ors* [2006] NSWLEC 720, para. 98.

political questions lie outside of Article III jurisdiction of federal courts'[409] (see Section 4.2.1.1 above). The political question doctrine aims at limiting judicial interference with the activities of the legislative and executive branches.[410] According to the political question doctrine, the judiciary cannot intervene in policy issues as those are to be decided by the democratically elected branches of the US government. Its genesis lies in the principles of the separation of powers and it was initially articulated by Chief Justice Marshall in the early nineteenth century.[411] The doctrine is grounded in 'the courts' reluctance to invade the constitutionally allocated powers of the executive and legislative branches of government'[412] and, as a result, it has been at the forefront of the courts' rejection of a number of climate change claims in the US and Canada. Six categories of non-justiciable suits were specified in *Baker v. Carr*:

> Prominent on the surface of any case held to involve a political question is found [(1)] a textually demonstrable constitutional commitment of the issue to a coordinate political department; or [(2)] a lack of judicially discoverable and manageable standards for resolving it; or [(3)] the impossibility of deciding without an initial policy determination of a kind clearly for nonjudicial discretion; or [(4)] the impossibility of a court's undertaking independent resolution without expressing lack of respect due coordinate branches of government; or [(5)] an unusual need for unquestioning adherence to a political decision already made; or [(6)] the potentiality of embarrassment from multifarious pronouncements by various departments on one question.[413]

It has been noted that the political question doctrine has at times turned into an excuse 'to evade the courts' responsibility to decide serious justiciable issues in environmental law,' which has been used as 'an unwarranted escape hatch' thwarting 'effective judicial redress for environmental harms.'[414] According to Daly, a more accurate description of the political question doctrine would be 'an attitude of judicial restraint, adopted by judges when they are asked to review certain categories of sensitive decision.'[415]

Political questions proved to be an insurmountable obstacle in the Canadian *Friends of the Earth v. Minister of the Environment* claim. In this case, the

---

[409] *Native Village of Kivalina v. ExxonMobil Corp.*, 663 F.Supp.2d 863 (N.D.Cal. 2009), p. 4 citing *Corrie v. Caterpillar, Inc.*, 503 F.3d 974 (9th Cir.2007).

[410] B.C. Mank, 'Standing and global warming: is injury to all injury to none?', 35 *Envtl. L.* 1, 2005, p. 29.

[411] P. Weinberg, ''Political questions': an invasive species infecting the courts,' 19 *Duke Envtl. L. & Pol'y F.* 155, 2008, p. 156, citing *Marbury v. Madison,* 5 U.S. (1 Cranch) 137, 166 (1803).

[412] Weinberg 2008, pp. 156-157.

[413] *Baker v. Carr*, 369 U.S. 186, 226 (1962), p. 217.

[414] Weinberg 2008, p. 155.

[415] P. Daly, 'Justiciability and the 'political question' doctrine,' *P.L.* 2010, Jan, 160-178, p. 176.

applicant brought three applications for judicial review seeking declaratory and mandatory relief attempting to ensure the government's compliance with the Kyoto Protocol. The first application concerned the Minister's alleged failure to comply with the duty to prepare an initial Climate Change Plan that fulfilled Canada's Article 3(1) obligations under the Kyoto Protocol. The second application for judicial review alleged that the Governor in Council failed to publish proposed regulations in the Canada Gazette and to prepare a statement setting out the GHG emissions reduction reasonably expected to result from each proposed regulatory change. The third ground for review alleged that the Governor in Council failed to amend or repeal regulations necessary to ensure Canada's compliance with its obligations under Article 3(1) of the Kyoto Protocol.[416] The court observed that the Minister's initial Climate Change Plan made it very clear that the Canadian government had no intention to meet its Kyoto commitments[417] noting that Canada's emissions had been steadily growing since 1990. The respondents contended that the issues raised by the applicant were not justiciable because they involved public policy and legislative choices that were not up to the court to make.[418] In this case the standing of the applicants was only challenged on the ground of non-justiciability; the court was satisfied that otherwise the applicant had met the requirements of public interest standing in that it had 'a genuine interest in the subject matter raised,' that there was 'a serious issue presented,' and that there was 'no other reasonable and effective way to bring these matters before the Court.'[419] The court noted that the parties agreed about the principles of justiciability; they only differed in their views on the application of those principles. It was agreed that even 'a largely political question' could be judicially reviewed if it 'possess[ed] a sufficient legal component to warrant a decision by a court.'[420] The court found that the applications in question were policy-laden and did not contain 'a sufficient legal component to permit judicial review.'[421]

---

[416] *Friends of the Earth – Les Ami(e)s de la Terre v. The Governor in Council and The Minister of the Environment*, 2008 FC 1183, 20 October 2008, paras 3-5.

[417] *Friends of the Earth – Les Ami(e)s de la Terre v. The Governor in Council and The Minister of the Environment*, 2008 FC 1183, 20 October 2008, para. 11.

[418] *Friends of the Earth – Les Ami(e)s de la Terre v. The Governor in Council and The Minister of the Environment*, 2008 FC 1183, 20 October 2008, para. 7.

[419] *Friends of the Earth – Les Ami(e)s de la Terre v. The Governor in Council and The Minister of the Environment*, 2008 FC 1183, 20 October 2008, para. 20.

[420] *Friends of the Earth – Les Ami(e)s de la Terre v. The Governor in Council and The Minister of the Environment*, 2008 FC 1183, 20 October 2008, para. 24 quoting *Reference Re Canada Assistance Plan (B.C.)*, [1991] 2 S.C.R. 525, para. 27, 83 D.L.R. (4th) 297.

[421] *Friends of the Earth – Les Ami(e)s de la Terre v. The Governor in Council and The Minister of the Environment*, 2008 FC 1183, 20 October 2008, para. 24.

[36] Furthermore, if the Court is not permitted by the principles of justiciability to examine the substantive merits of a Climate Change Plan that dubiously claimed Kyoto compliance, it would be incongruous for the Court to be able to order the Minister to prepare a compliant Plan where he has deliberately and transparently declined to do so for reasons of public policy.[422]

The court dismissed all the applications concluding that it had 'no role to play reviewing the reasonableness of the government's response to Canada's Kyoto commitments.'[423]

In the US, *Connecticut v. American Electric Power* 2005, too, was rejected in first instance on the grounds of non-justiciability of political questions. In this case, eight states,[424] the New York City, and three land trusts[425] instituted a public nuisance global warming action against the five largest GHG emitting companies among electricity generators.[426] The plaintiffs alleged, *inter alia*, that the defendants' carbon dioxide emissions contributed to global warming, which caused 'irreparable harm' to their property and threatened the 'health, safety, and well-being of [...] citizens, residents, and environment.'[427] The plaintiffs asserted that the natural processes that removed GHGs from the atmosphere were 'unable to keep pace with the level of carbon dioxide emissions' and that the defendants' past, present, and future emissions would 'remain in the atmosphere and contribute to global warming for many decades and, possibly, centuries.' The plaintiffs anticipated that global warming would have 'severe adverse impacts in the United States.'[428] They asked the court to find the defendants jointly and severally liable for their contribution to the 'ongoing public nuisance' of global warming and sought an injunction requiring the defendants to reduce their emissions 'by a specified percentage each year for at least a decade.'[429] The court granted the defendants' motion to dismiss on the grounds of the non-justiciability-of-political-

---

[422] *Friends of the Earth – Les Ami(e)s de la Terre v. The Governor in Council and The Minister of the Environment*, 2008 FC 1183, 20 October 2008.

[423] *Friends of the Earth – Les Ami(e)s de la Terre v. The Governor in Council and The Minister of the Environment*, 2008 FC 1183, 20 October 2008, para. 46.

[424] The states of Connecticut, New York, California, Iowa, New Jersey, Rhode Island, Vermont, and Wisconsin.

[425] The Open Space Institute, Inc.; the Open Space Conservancy, Inc.; and the Audubon Society of New Hampshire.

[426] American Electric Power Company, Inc.; American Electric Power Service Corporation; the Southern Company; Tennessee Valley Authority; Xcel Energy, Inc.; and Cinergy Corporation.

[427] *Connecticut v. American Electric Power Co., Inc.*, 406 F.Supp. 2d 265 (S.D.N.Y. 2005), 35 Envtl. L. Rep. 20, 186, p. 268.

[428] *Connecticut v. American Electric Power Co., Inc.*, 406 F.Supp. 2d 265 (S.D.N.Y. 2005), 35 Envtl. L. Rep. 20, 186, pp. 268-269.

[429] *Connecticut v. American Electric Power Co., Inc.*, 406 F.Supp. 2d 265 (S.D.N.Y. 2005), 35 Envtl. L. Rep. 20, 186, p. 270.

questions doctrine. The court found it impossible to resolve the plaintiffs' claims without making an initial policy determination, which, it pointed out, was the prerogative of the elected branches of the government. On standing and its causation and redressability elements in particular, the court stated: '[t]he extraordinary allegations and relief sought in this case render it one in which an analysis of Plaintiffs' standing would involve an analysis of the merits of Plaintiffs' claims.'[430] On appeal, the US Court of Appeals for the Second Circuit vacated the judgment of the district court and the case was remanded for further proceedings. The appellate court held that the plaintiffs-appellants' claims did not present non-justiciable political questions and that the district court had erred in dismissing the complaints on those grounds.[431]

4.2.2.2 Standing

In injunctive relief claims, the preconditions for plaintiffs' standing are more rigorous than those in cases involving procedural injury where standing requirements are relaxed (see Section 4.2.1.1). Several actions for injunctions failed because plaintiffs were unable to show that they had standing to bring their claims.

Although in *Center for Biological Diversity v. Abraham*, the court held that the plaintiffs did have standing, its decision was not based on the plaintiffs' global warming claims. In this case, three environmental organizations[432] brought an action against eighteen government agencies[433] seeking declaratory and injunctive relief. According to the plaintiffs, the defendants violated the 1992 Energy Policy Act provisions governing acquisition of alternative fuel vehicles (AFV) and publication of compliance reports. The plaintiffs argued that the defendants' actions contributed to global warming, which, in turn, caused harm to the plaintiffs. To satisfy the constitutional standing requirements, the plaintiffs had to demonstrate that: (1) they had suffered an injury in fact that was (a) concrete and particularized

---

[430] *Connecticut v. American Electric Power Co., Inc.*, 406 F.Supp. 2d 265 (S.D.N.Y. 2005), 35 Envtl. L. Rep. 20, 186, p. 271, footnote 6.

[431] In 2011, the Supreme Court reversed the appellate court's judgment ruling that the CAA, and the EPA action authorized by it, displaced any federal common-law right to seek abatement of GHG emissions from fossil-fuel fired power plants, see *American Electric Power Co., Inc. v. Connecticut*, 131 S.Ct. 2527 U.S., 2011.

[432] Center for Biological Diversity, Bluewater Network, and Sierra Club.

[433] The defendants, all sued in their official capacities, are the secretaries of Energy; Commerce; Defense; Interior; Veteran Affairs; Transportation; Agriculture; Health and Human Services; Housing and Urban Development; Labor; State; and Treasury; the postmaster general of the US Postal Service; the administrators of the National Aeronautics and Space Administration; the US Environmental Protection Agency; and the General Services Administration; the chair of the US Nuclear Regulatory Commission; and the US Attorney General. The plaintiffs later agreed to dismiss the postmaster general and the US Postal Service as defendants.

and (b) actual or imminent; that (2) the injury was fairly traceable to the challenged action of the defendants; and that (3) it was likely that the injury would be addressed by a favourable decision. The court granted plaintiffs standing but for reasons other than their climate change claims. The plaintiffs' claims regarding aesthetic injuries and the adverse health effects of smog and air pollution caused by the defendants' conduct were found to satisfy the injury element of standing.[434] The court noted that the concerns presented with regard to global warming were 'too general, too unsubstantiated, too unlikely to be caused by defendants' conduct, and/or too unlikely to be redressed by the relief sought to confer standing.'[435] The court granted declaratory and injunctive relief with regard to Count II (the defendants violated the Energy Policy Act in that they failed to compile and to properly make publicly available the compliance reports required by the Act) and Count III (the defendants violated the Energy Policy Act in that the Department of Energy did not meet the rulemaking deadlines for private and local fleets provided in the Act) and only declaratory relief with regard to Count I (the defendants violated the Energy Policy Act in that several among them did not meet the annual AFV-acquisition requirements set forth in the Act).[436]

In *Korsinsky v. EPA*, the plaintiff was found to have failed to establish Article III standing. In this case, the plaintiff, Mr Gersh Korsinsky, appearing in court for himself, filed a public nuisance complaint against the US Environmental Protection Agency (EPA), the NYS Department of Environmental Conservation, and the NYC Department of Environmental Protection. The plaintiff alleged that the defendants contributed to global warming by emitting approximately 6500 million tons of carbon dioxide annually and failing to implement measures that would reduce GHG emissions. The plaintiff requested that the defendants be held jointly and severally liable for their contributions to global warming and sought an injunction enjoining each defendant from contributing further to global warming by 'eliminating its emissions of carbon dioxide,' and an order requiring the defendants to use a scientific device of his own invention.[437] The plaintiff complained that he was more vulnerable to disease-causing environmental pollution because he had suffered from various allergies all his life and that he had developed a mental sickness from learning of the danger of pollution. The court found that the plaintiff alleged no injury that could be redressed by a favourable decision. Thus, the plaintiff did not fulfil the requirements for standing (an injury in fact, causation, and redressability).

Other plaintiffs have managed to overcome the challenges associated with demonstrating standing. Standing has been successfully established in *Northwest*

---

[434] *Center for Biological Diversity v. Abraham*, 218 F.Supp. 2d 1143 (N.D.Cal. 2002), p. 1155.
[435] *Center for Biological Diversity v. Abraham*, 218 F.Supp. 2d 1143 (N.D.Cal. 2002), p. 1155.
[436] *Center for Biological Diversity v. Abraham*, 218 F.Supp. 2d 1143 (N.D.Cal. 2002).
[437] *Korsinsky v. EPA*, 2005 WL 2414744 (S.D.N.Y.).

Environmental Defense Center v. Owens Corning. Environmental organizations[438] sued the Owens Corning Corporation for building a facility with a potential to emit over 250 tons of GHGs annually without having obtained a preconstruction permit required under the CAA. The plaintiffs alleged that emissions from the defendant's facility would contribute to global warming, which, in turn, would 'harm environmental resources in Oregon used or enjoyed by members of the Plaintiff organizations.'[439] The plaintiffs sought declaratory and injunctive relief. The plaintiffs were found to have standing to sue. In order to establish federal jurisdiction, they had to satisfy the Article III standing requirements and demonstrate: (1) that they suffered or were about to suffer an injury in fact ('an invasion of a legally protected interest that is (a) concrete and particularized, and (b) actual or imminent, not conjectural or hypothetical'); (2) a causal connection between the injury and the conduct; and (3) redressability (the injury would be redressed by a favourable decision).[440] The court found that the plaintiffs satisfied all the elements of the standing inquiry. It noted that issues such as global warming and ozone depletion may be of 'wide public significance' but they are neither 'abstract questions' nor mere 'generalized grievances.' The court further stated that an injury 'is not beyond the reach of the courts simply because it is widespread.'[441] The court also distinguished the case from *Connecticut v. American Electric Power Company* 2005,[442] in which the district court declined to resolve the plaintiffs' claims relying on the political question doctrine, and stated the following:

> At issue here is nothing more than whether the courts will enforce the Congressional mandate set forth in the Clean Air Act and its enabling regulations. This court is not being asked to make a free-wheeling policy choice and decide whether global warming is, or is not, a serious threat or what measures should be taken to remedy that problem.[443]

Thus, the court did not rely on the political question doctrine and made a crucial finding that a global warming plaintiff had standing to sue under the CAA.

---

[438] Northwest Environmental Defense Center, Oregon Center for Environmental Health, and Sierra Club.

[439] *Northwest Environmental Defense Center v. Owens Corning Corp.,* 434 F.Supp. 2d 957 (D.Or. 2006), p. 960.

[440] *Northwest Environmental Defense Center v. Owens Corning Corp.,* 434 F.Supp. 2d 957 (D.Or. 2006), p. 962.

[441] *Northwest Environmental Defense Center v. Owens Corning Corp.,* 434 F.Supp. 2d 957 (D.Or. 2006), p. 969.

[442] *Connecticut v. American Electric Power Co., Inc.,* 406 F.Supp. 2d 265 (S.D.N.Y. 2005), 35 Envtl. L. Rep. 20, 186.

[443] *Northwest Environmental Defense Center v. Owens Corning Corp.,* 434 F.Supp. 2d 957 (D.Or. 2006), p. 970.

In *Massachusetts v. EPA*, a case that is generally accepted as one of the most important global warming decisions in America and worldwide,[444] the US Supreme Court held that plaintiffs had standing to bring their claim. The case was decided in 2007; however, it goes back to the late 1990s when the International Center for Technology Assessment together with several other private organizations petitioned the Environmental Protection Agency to regulate GHGs from new motor vehicles under the Clean Air Act (CAA). In 2003, the EPA denied the rulemaking petition stating that it had no authority under the CAA to regulate GHGs and that even if it had such an authority, it would not exercise it. The petitioners, now joined by intervener States and local governments,[445] sought review of the EPA's decision in the US Court of Appeals for the DC Circuit. The Court denied the petition for review. The petitioners appealed the decision, and the Supreme Court entered the climate change debate for the first time. The Supreme Court found in favour of the petitioners holding that the EPA had the authority under the CAA to regulate GHGs and that its refusal to do so was arbitrary and capricious. Notably, the Court found that Massachusetts had standing to challenge the federal agency's decision. The court noted the 'special solicitude' afforded to states regarding their procedural right for standing. Additionally, Massachusetts satisfied the traditional requirements for standing, *i.e.* injury, causation, and redressability. The court determined that climate change had caused Massachusetts a concrete and particularized injury notwithstanding the fact that climate-change risks were 'widely shared.'[446] The court recognized that 'predicted increases in greenhouse gas

---

[444] *Massachusetts v. EPA*, 127 S.Ct. 1438, U.S., 2007. On the case see, *e.g.* Abate, 2008; T.J. Boutrous Jr. & D. Lanza, 'Global warming tort litigation: the real "public nuisance,"' Vol. 35:80 *Ecology Law Currents,* 2008; M. Eschen Mills, 'The global warming case: Massachusetts v. Environmental Protection Agency,' 7 *Sustainable Dev. L. & Pol'y* 67, 2007; L. Heinzerling, 'Climate change in the Supreme Court,' 38 *Envtl. L.* 1, 2008; J.R. Pidot, 'Global warming in the courts: the *Massachusetts v. EPA* decision and its implications' (online paper), Georgetown Environmental Law & Policy Institute, Georgetown University Law Center, April 2007, available from:
<www.law.georgetown.edu/gelpi/current_research/documents/GWL_Update_4.24.07.pdf> (last visited on 14 April 2009); D. Stevenson, 'Special solicitude for state standing: Massachusetts v. EPA,' 112 *Penn St. L. Rev.* 1, 2007; Sugar 2007. Let us also note that even before the Supreme Court's decision in *Massachusetts,* in *Border Power Plant Working Group v. Department of Energy* (260 F.Supp. 2d 997 (S.D.Cal. 2003)), the court held that the Department's environmental analysis was inadequate because it failed to address the 'potential environmental impacts' of carbon dioxide emissions, despite CO_2 not having been classified by the EPA as a 'criteria pollutant.'

[445] States of California, Connecticut, Illinois, Maine, Massachusetts, New Jersey, New Mexico, New York, Oregon, Rhode Island, Vermont, and Washington; local governments of District of Columbia, American Samoa, New York City, and Baltimore; and the following NGOs: Center for Biological Diversity, Center for Food Safety, Conservation Law Foundation, Environmental Advocates, Environmental Defense, Friends of the Earth, Greenpeace, International Center for Technology Assessment, National Environmental Trust, Natural Resources Defense Council, Sierra Club, Union of Concerned Scientists, and US Public Interest Research Group.

[446] *Massachusetts v. EPA*, 127 S.Ct. 1438, U.S., 2007, p. 522.

emissions from developing nations, particularly China and India, [were] likely to offset any marginal domestic decrease' in GHG emissions.[447] It found, however, that GHG emissions from motor vehicles in the US made a 'meaningful contribution' to global warming, which, in turn, affected Massachusetts. The Court held that a 'reduction in domestic emissions would slow the pace of global emissions increases, no matter what happens elsewhere.'[448] Thus, Massachusetts' injuries caused by global warming would be redressed by the remedy sought by the petitioners.

The Supreme Court's decision in *Massachusetts v. EPA* has been hailed as ground-breaking and a major success for climate change plaintiffs.[449] Yet, it has been noted that the Court's reference to the special solicitude afforded to Massachusetts as a state may heighten the standing threshold for private plaintiffs.[450] The Supreme Court's analysis of standing has been criticized for creating 'a two-tiered system separating states from other petitioners for review.'[451] First, the Court found that states had 'special solicitude' to sue due to their quasi-sovereign interest and, second, the Supreme Court also held that Massachusetts had standing under the traditional Article III requirements. Some authors have found this approach perplexing, and the 'special solicitude' distinction has been said to grant states a favoured status resulting in negative implications for the standing of individuals in climate change cases.[452]

Having been found to have the statutory authority to regulate GHGs from new motor vehicles under the CAA, the EPA has been taking steps to regulate GHGs and formulate GHG emissions rules.[453] In the wake of *Massachusetts v. EPA*, the Agency has also commenced rulemaking to set limits on GHG emissions from new, modified, and existing fossil-fuelled power plants and committed to issuing a final rule by May 2012 (still in progress at the time of writing).

Procedural standing was successfully established in *Center for Biological Diversity v. Brennan*[454] – an action for declaratory and injunctive relief to declare the defendants in violation of the Global Change Research Act 1990 (GCRA) and to compel them to issue the Research Plan and Scientific Assessment as directed by statute.[455] The defendants contended that the plaintiffs[456] lacked standing to sue.

---

[447] *Massachusetts v. EPA*, 127 S.Ct. 1438, U.S., 2007, p. 524.

[448] *Massachusetts v. EPA*, 127 S.Ct. 1438, U.S., 2007, p. 526.

[449] Pidot 2007; Smith & Shearman 2006, p. 64.

[450] Abate 2008, pp. 147-148; Stevenson 2007, p. 74; Sugar 2007, pp. 542-544.

[451] Sugar 2007, p. 541.

[452] Heinzerling 2008, p. 16; Sugar 2007, p. 544; on the favoured position of states and Attorneys General, see Stevenson 2007.

[453] See <http://epa.gov/climatechange/EPAactivities/regulatory-initiatives.html> (last visited on 10 July 2012).

[454] *Center for Biological Diversity v. Brennan*, 571 F.Supp. 2d 1105 (N.D.Cal., 2007).

[455] *Center for Biological Diversity v. Brennan*, 571 F.Supp. 2d 1105 (N.D.Cal., 2007), pp. 1112-1113.

[456] Center for Biological Diversity, Greenpeace, Inc., and Friends of the Earth.

The court found that the plaintiffs had adequately alleged a procedural injury with respect to the Scientific Assessment and the Research Plan. The plaintiffs' concrete interests were also found to fall within the zone of interests protected by the GCRA. As in other cases mentioned before, requirements for procedural standing were relaxed, and the traceability and redressability prongs were likewise found to have been successfully demonstrated. On the merits, the court found that the defendants had unlawfully withheld action required by the GCRA and ordered an injunction requiring the defendants to comply with the statute and to issue a revised Research Plan and Scientific Assessment.

In *Connecticut v. American Electric Power* 2009, the appellate court found that the plaintiff states had *parens patriae* standing (suing on behalf of their citizens) in their quasi-sovereign capacity. The court further held that in their proprietary capacity, they, along with the land trusts, also had Article III standing to bring their claim. The court held that current and future injury-in-fact had been sufficiently alleged. It further held that the injury was fairly traceable to the defendants' actions, thus finding the causation element sufficiently demonstrated. With regard to the defendants' argument that they could not be solely responsible for the 'collective effect of worldwide emissions' allegedly causing the injury, the court noted that that was an issue 'best left to the rigors of evidentiary proof at a future stage of the proceedings, rather than dispensed with as a threshold question of constitutional standing.'[457] As regards the Article III redressability requirement, the judges pointed out the following:

> [In *Massachusetts v. EPA*] [t]he court recognized that regulation of motor vehicle emissions would not 'by itself *reverse* global warming,' but that it was sufficient for the redressability inquiry to show that the requested remedy would '*slow* or *reduce* it.' […] In other words, that courts could provide some measure of relief would suffice to show redressability, and the proposed remedy need not address or prevent all harm from a variety of sources.[458]

4.2.2.3 Causation

In injunctive and/or declaratory relief actions, the courts have only explicitly addressed causation as part of the standing analysis, which, as discussed earlier, is characterized by a lower standard of proof. One may speculate, however, that the sheer complexity of the causal chain between climate change-related damage and

---

[457] *Connecticut v. American Electric Power Co., Inc.,* 582 F.3d 309, C.A.2 (N.Y.), 2009, p. 61.

[458] *Connecticut v. American Electric Power Co., Inc.,* 582 F.3d 309, C.A.2 (N.Y.), 2009, p. 63, emphasis in the original.

GHG emissions from a particular source would pose a major challenge if the court were to engage in an analysis of the merits of a claim. A link would have to be established between: (1) GHG emissions from a given source; (2) climate change; (3) manifestations of climate change, such as the rising sea levels or extreme weather events; and, ultimately, (4) damage.

Causation challenges have been emphasized in *Center for Biological Diversity v. Abraham* whereby the court found that the plaintiffs' global warming claims were 'too unlikely to be caused by defendants' conduct.'[459] It is significant that in finding that the plaintiffs had standing to bring their claim, the court did not base its reasoning on the plaintiffs' climate change complaints.

In *Korsinsky v. EPA*, the plaintiff was unable to establish the causation element of standing.[460]

Yet, causation as an element of standing was successfully shown in *Northwest Environmental Defense Center v. Owens Corning* case. In its standing analysis, the court reasoned that there was a causal link between the emissions from the defendant's facility and the harm caused to the plaintiffs by the resultant climate change:

> While Defendant is not the sole entity allegedly discharging pollutants into the atmosphere that may adversely impact the Plaintiffs, the 'fairly traceable' element does not require that a plaintiff show to a scientific certainty that the defendant's emissions, and only the defendant's emissions, are the source of the threatened harm. [...] It is sufficient for Plaintiffs to assert that emissions from Defendant's facility will contribute to the pollution that threatens Plaintiffs' interests.[461]

The court's reference to scientific certainty is of interest here as it has been suggested that '[e]nhanced certainty regarding the scientific data [...] has made injury and causation determinations easier'[462] and as scientific certainty keeps growing, proof of causation should become accordingly easier. It is not yet clear whether because of growing scientific certainty determinations of causation would likewise be easier to make on the merits. Since the standard of proof will necessarily be higher at the merits stage, increased scientific certainty concerning the link between anthropogenic emissions and climate change alone will not suffice. Plaintiffs may need to demonstrate to what extent their injury can be considered to have been caused by defendants, in which case apportionment and quantification of liability may present considerable difficulties.

---

[459] *Center for Biological Diversity v. Abraham*, 218 F.Supp. 2d 1143 (N.D.Cal. 2002), paras 6-7.

[460] *Korsinsky v. EPA*, 2005 WL 2414744 (S.D.N.Y.).

[461] *Northwest Environmental Defense Center v. Owens Corning Corp.*, 434 F.Supp. 2d 957 (D.Or. 2006), p. 967.

[462] Abate 2008, pp. 171-172.

Bearing in mind the relaxed requirements for procedural standing, the plaintiffs in *Massachusetts v. EPA* and *Center for Biological Diversity v. Brennan* were able to satisfy the causation requirement for standing. In *Center for Biological Diversity v. Brennan*, the court found that 'the plaintiffs' procedural and informational injuries [could] be directly traced to the defendants' failure to issue a revised Research Plan and Scientific Assessment.' [463] In *Massachusetts v. EPA*, the Supreme Court assumed a much-criticized two-tiered approach to standing (see Section 4.2.2.2) whereby Massachusetts was found to have a procedural right for standing due to 'special solicitude' afforded to it as a state. Further, the court held that Massachusetts satisfied all the three elements of Article III standing, including causation: GHGs emitted by vehicles that the EPA refused to regulate were found to play a major role in causing global warming, which, in turn, caused harm to Massachusetts.

Accordingly, in the *Connecticut v. American Electric Power* appeal 2009, the Second Circuit also found that the plaintiff-appellants' harm was 'fairly traceable' to the defendants' actions. The court noted, however, that the defendants' assertions to the effect that their contribution to the worldwide emissions did not allege causation were 'best left to the rigors of evidentiary proof at a future stage of the proceedings.' [464] This seems to suggest that even though causation may be successfully proven as an element of Article III standing, considerable evidentiary challenges are to be expected at the merits stage.

The *Friends of the Earth v. Minister of the Environment* case did not address causation as the Canadian court found it non-justiciable.

4.2.2.4 Attribution

It has already been mentioned that the problem of attribution of harm to defendants has not been extensively addressed by domestic courts (see Section 4.2.1.3). Attribution is an issue to be considered on the merits, and injunctive and/or declaratory relief claims discussed in this chapter have not reached that stage of the proceedings.

Yet, in the *Connecticut v. American Electric Power* appeal 2009, the defendants' contention that 'many others contribute to global warming in a variety of ways' and that 'only the collective effect of worldwide emissions allegedly causes injury'[465] raised the question of attribution. In leaving this issue for the merits stage, the Second Circuit noted:

---

[463] *Center for Biological Diversity v. Brennan,* 571 F.Supp. 2d 1105 (N.D.Cal. 2007), p. 1122.

[464] *Connecticut v. American Electric Power Co., Inc.,* 582 F.3d 309, C.A.2 (N.Y.), 2009, p. 61.

[465] *Connecticut v. American Electric Power Co., Inc.,* 582 F.3d 309, C.A.2 (N.Y.), 2009, p. 61.

Plaintiffs have sufficiently alleged that their current and future injuries are 'fairly traceable' to Defendants' conduct. For purposes of Article III standing *they are not required to pinpoint which specific harms of the many injuries they assert are caused by particular Defendants, nor are they required to show that Defendants' emissions alone cause their injuries*. It is sufficient that they allege that Defendants' emissions contribute to their injuries.[466]

As discussed in Section 4.2.1.3, attribution of harm to a particular defendant in view of the cumulative nature of climate change is likely to be problematic.

4.2.3   Claims for Compensation

In some instances, climate change cases have been brought as claims for compensation, or, as they are known in common law, actions for damages. Compensation claims constitute the least numerous category of climate change cases. Although generally unsuccessful regardless of some commentators' optimism,[467] claims for compensation play an important role as, unlike claims related to procedural injury or injunctive relief claims, they are relevant to situations when adaptation to, and remediation of, the already happening climate change is at stake. All the compensation claims adjudicated thus far worldwide have been brought in the US. The US remains the only developed country without a mitigation policy based on the 1997 Kyoto Protocol and, given its common law tradition, it is not surprising that it is in the US that litigants have resorted to actions for damages. It is also important to note that whether or not climate change defendants are eventually found liable for damages, American practice may provide some important lessons for other jurisdictions.

4.2.3.1 Political Question Doctrine

Legal issues involved in actions for damages have largely been similar to the ones discussed in relation to injunctive relief cases (see Sections 4.2.1.1-4.2.2.4). Non-justiciability of political questions has been particularly problematic. For example, *California v. General Motors Corporation*, a public nuisance global warming

---

[466] *Connecticut v. American Electric Power Co., Inc.,* 582 F.3d 309, C.A.2 (N.Y.), 2009, p. 61, emphasis added.

[467] *E.g.* D.A. Grossman, 'Warming up to a not-so-radical idea: tort-based climate change litigation,' 28 *Colum. Envtl. L.* 1, 3, 2003, p. 58 (it may be possible 'to establish the basis for a damage award in a public nuisance suit'); Hsu 2008; Posner 2007, p. 1928 (claims made in US domestic courts that GHG emissions leading to climate change violate human rights may result in American courts awarding damages to victims).

lawsuit for damages, filed by the State of California against several automobile manufacturers[468] was rejected on the basis of the political question doctrine. The plaintiff in this case sought declaratory judgment for 'future monetary expenses and damages incurred by the State of California in connection with the nuisance of global warming.'[469] The plaintiff alleged a number of global-warming injuries caused by GHG emissions from motor vehicles manufactured by the defendants. The plaintiff pointed out that the scientific debate over global warming was 'over' and that there was 'a clear scientific consensus' that global warming had begun and that most of it was caused by GHG emissions, primarily carbon dioxide from fossil fuel combustion. According to the plaintiff, global warming caused an increase in the winter average temperatures in the Sierra Nevada region and thus, a reduction in the snow pack – a major water source in California. The plaintiff noted an increased risk of flooding due to the rising sea levels, increased coastline erosion, increased duration and frequency of heat waves, and increases in the risk and intensity of wildfires.[470] The court dismissed the case on the grounds of non-justiciability of political questions. It held that (a) resolution of the plaintiff's federal common law nuisance claim would require it to make 'an initial policy decision' as the claims touched upon public and foreign policy. The court ruled that it could not adjudicate the plaintiff's claim 'without making an initial policy determination of a kind clearly for nonjudicial discretion.' The court noted that (b) the plaintiff's claim implicated 'a textually demonstrable constitutional commitment to the political branches.' It stated that (c) it lacked 'judicially discoverable or manageable standards' by which to resolve the plaintiff's claim.[471]

The *Kivalina* claim, too, was found to be non-justiciable. The case involved a suit filed by the residents of the Alaskan Native Village of Kivalina against a number of oil and electric utility industries[472] to recover damages from the public nuisance of global warming allegedly caused by the defendants. The plaintiffs claimed that global warming was destroying their village and soon the village would have to be abandoned and its population relocated, the cost of which had been estimated at 400 million US dollars. According to the claimants,

---

[468] General Motors Corp.; Toyota Motor North America, Inc.; Ford Motor Co., American; Honda Motor Co., Inc.; Daimler Chrysler Corp.; and Nissan North America, Inc.

[469] *California v. General Motors Corp.,* 2007 WL 2726871 (N.D.Cal.), p. 2.

[470] *California v. General Motors Corp.,* 2007 WL 2726871 (N.D.Cal.), pp. 1-2.

[471] *California v. General Motors Corp.,* 2007 WL 2726871 (N.D.Cal.), pp. 6-16.

[472] ExxonMobil Corporation; BP P.L.C.; BP America, Inc.; BP Products North America, Inc.; Chevron Corporation; Chevron USA, Inc.; Conocophillips Company; Royal Dutch Shell PLC; Shell Oil Company; Peabody Energy Corporation; the AES Corporation; American Electric Power Company, Inc.; American Electric Power Services Corporation; DTE Energy Company; Duke Energy Corporation; Dynegy Holdings, Inc.; Edison International; MidAmerican Energy Holdings Company; Mirant Corporation; NRG Energy; Pinnacle West Capital Corporation; Reliant Energy, Inc.; The Southern Company; and XCEL Energy, Inc.

16. Global warming has severely harmed Kivalina by reducing the sea ice commonly present in the fall, winter and spring at Kivalina. The sea ice – particularly land-fast sea ice – acts as a protective barrier to the coastal storms that batter the coast of the Chukchi Sea. Due to global warming, the sea ice forms later in the year, attaches to the coast later, breaks up earlier, and is less extensive and thinner, thus subjecting Kivalina to coastal storm waves and surges. These storms and waves are destroying the land upon which Kivalina is located.

17. Impacts of global warming have damaged Kivalina to such a grave degree that Kivalina is becoming uninhabitable and must now relocate its entire community.

According to the plaintiffs, the defendants contributed to global warming 'through their emissions of large quantities of greenhouse gases' and they had knowingly done so for many years. The plaintiffs also alleged a conspiracy to suppress the awareness of the link between the defendants' emissions and global warming.[473] The plaintiffs requested the court to hold each defendant 'jointly and severally liable for creating, contributing to, and maintaining a public nuisance' of global warming, hold conspiracy defendants 'jointly and severally liable for civil conspiracy,' and hold each defendant 'jointly and severally liable for concert of action.' On that basis, the plaintiffs requested monetary damages and declaratory relief for 'future monetary expenses and damages as may be incurred' by the plaintiffs in connection with global warming.[474] The court found the plaintiffs' claim non-justiciable under the political question doctrine due to lack of judicially discoverable and manageable standards and absence of an initial policy determination. It noted that 'allocation of fault – and cost – of global warming [was] a matter appropriately left for determination by the executive or legislative branch in the first instance.'[475]

Non-justiciability of political questions also proved a contentious point in *Comer*. This global warming public nuisance class action suit was originally filed against US property insurers in the wake of the destruction brought by Hurricane Katrina in August 2005. Before long, the claim was amended as to include petrochemical companies as defendants. The plaintiffs alleged that Hurricane Katrina 'evolved into a storm of unprecedented strength and destruction' due to

---

[473] 2008 WL 594713 (N.D.Cal.) (Trial Pleading) Complaint for Damages Demand for Jury Trial (26 February 2008), *Native Village of Kivalina v. ExxonMobil Corp.*, paras 3-5.
[474] Relief requested in 2008 WL 594713 (N.D.Cal.) (Trial Pleading) Complaint for Damages Demand for Jury Trial (26 February 2008), *Native Village of Kivalina v. ExxonMobil Corp.*
[475] *Native Village of Kivalina v. ExxonMobil Corp.*, 663 F.Supp.2d 863 (N.D.Cal. 2009), p. 10; the decision of the district court was confirmed on appeal, see *Native Village of Kivalina v. ExxonMobil Corp.*, 2012 WL 4215921 (9th Cir.(Cal.)) (21 September 2012) (NO. 09-17490).

global warming caused by the activities' of the defendant oil companies.[476] The district court acknowledged that, in its complexity, the case was unmanageable. It dismissed the plaintiffs' causes of action against insurance companies and mortgage lenders and permitted to file a third amended complaint.[477] Ned Comer and thirteen other individuals brought an action for damages against public utilities, power companies, and coal-mining and chemical manufacturing companies. The plaintiffs' causes of action included nuisance, negligence, unjust enrichment, civil conspiracy, fraudulent misrepresentation and concealment, and trespass. The central public nuisance claim was based on the defendants' alleged contribution to global warming leading to Katrina, a hurricane of unprecedented intensity, which caused damage to the plaintiffs. The plaintiffs' case was dismissed due to their lack of standing (see Section 4.2.3.2) and on the basis of non-justiciability of political questions.[478] On appeal, the Fifth Circuit reversed the judgment of the district court and remanded the cases for further proceedings. The Fifth Circuit found that, on the contrary, none of the above claims presented non-justiciable political questions stressing that 'the federal courts [were] not free to invoke the political question doctrine to abstain from deciding politically charged questions like this one, but must exercise their jurisdiction as defined by Congress whenever the question is not exclusively committed to another branch of the federal government.'[479] The defendants requested a rehearing, which was granted. The decision of the appellate court was vacated but a rehearing never took place due to a loss of quorum. Since the appeal could not be reinstated, in 2011, the plaintiffs filed another lawsuit in the district court against numerous oil, coal, electric, and chemical companies. In addition to concluding that the suit was barred under the doctrines of *res judicata* and collateral estoppel (*i.e.* the plaintiffs had already had an opportunity to litigate their claims), the court again found that the plaintiffs' claims presented non-justiciable political questions. The court stated that it could not decide on whether the defendants' emissions were 'unreasonable' because that would involve making an initial policy determination that 'had been entrusted to the EPA by Congress.'[480]

---

[476] 2005 WL 2913766 (Trial Pleading) First Amended Complaint (30 September 2005), *Comer v. Murphy Oil USA, Inc.*, paras 30-32.

[477] *Comer v. Nationwide Mutual Insurance Co.*, WL 1066645 (S.D.Miss), 23 February 2006; see also 2006 WL 1474089 (Trial Pleading) Third Amended Class Action Complaint (19 April 2006), *Comer v. Murphy Oil USA, Inc.*

[478] Order Granting Defendants' Motion to Dismiss, *Comer v. Murphy Oil USA, Inc.*, 2007 WL 6942285 (S.D.Miss.), 30 August 2007 (NO. CIVA 1:05-CV436LGRHW), para. 1.

[479] *Comer v. Murphy Oil USA, Inc.*, 585 F.3d 855, C.A.5 (Miss.), 2009, p. 13.

[480] *Comer v. Murphy Oil USA, Inc.*, Case 1:11-cv-00220-LG-RHW (S.D.Miss.), 20 March 2012.

4.2.3.2 Standing

Along with the challenges presented by the political question doctrine, standing in compensation claims has not been easy to determine. For instance, in *California v. General Motors Corporation*, the court explicitly distinguished the case before it from *Massachusetts v. EPA*, in which a federal state challenged federal agency action. Since the *California v. General Motors Corporation* action for damages was filed against private parties, *Massachusetts v. EPA* could not offer to the plaintiff a procedural right for standing because California was not challenging federal rule-making.[481]

The outcome of the *Kivalina* claim reaffirms the challenges posed by the standing threshold. *Kivalina* attracted a great deal of academic attention even before it was decided and many authors commented extensively on the plaintiffs' standing. With regard to the injury element of causation, Hsu has noted that 'the Inuit would be able to use the fact that they are one of the few groups that have already suffered some harm that is 'concrete and particularized,' and 'actual and imminent,' in the form of sinking villages due to softening permafrost.'[482] Mank has observed that 'an Alaska Native already harmed by global warming would have a much stronger case for injury than a plaintiff who can only allege general and common injuries from global warming.' Without a doubt, 'an Alaska Native whose village or home is destroyed by melting permafrost or coastal flooding could allege very specific injuries.'[483] The other two elements of Article III standing were seen as somewhat less clear-cut. And indeed, when finally decided, the claim failed on causation grounds. The court found that the plaintiffs had not established Article III standing failing to prove causation and observed:

> Although the 'traceability' of a plaintiff's harm to the defendant's actions need not rise to the level of proximate causation, Article III *does* require proof of a substantial likelihood that the defendant's conduct caused plaintiff's injury in fact.[484]

In *Comer v. Murphy Oil,* in addition to their claim being non-justiciable, the plaintiffs were found to have no standing to bring their claim because the harm they suffered was not traceable to individual defendants.[485] On appeal, however, the

---

[481] *California v. General Motors Corp.,* 2007 WL 2726871 (N.D.Cal.).

[482] Hsu 2008, p. 746.

[483] Mank 2005, pp. 80-81.

[484] *Native Village of Kivalina v. ExxonMobil Corp.,* 663 F.Supp.2d 863 (N.D.Cal. 2009), p. 10, italics in the original, citations omitted.

[485] Order Granting Defendants' Motion to Dismiss, *Comer v. Murphy Oil USA, Inc.,* 2007 WL 6942285 (S.D.Miss.) 30 August 2007 (NO. CIVA 1:05-CV436LGRHW).

court found that the plaintiffs-appellants did have standing 'to assert their public and private nuisance, trespass, and negligence claims,'[486] and not only on the basis of Mississippi's 'liberal standing requirements,' but also under Article III. The Fifth Circuit ruled that the plaintiffs had alleged 'actual, concrete injury in fact to their particular lands and property,' which could be 'redressed by the compensatory and punitive damages they [sought] for those injuries.'[487] When the district court heard the case in 2012, it again concluded that the plaintiffs had not successfully demonstrated Article III standing as they had not alleged injuries that were fairly traceable to the defendants' conduct. The court also distinguished the case before it from *Massachusetts v. EPA* and *Connecticut v. American Electric Power* discussed earlier. In those cases, the Supreme Court granted to federal states standing due to their sovereign status ('special solicitude') – a quality, the district court concluded, private citizens lacked.

4.2.3.3 Causation

As with claims for injunctive and/or declaratory relief, the question of causation in actions for damages has only been addressed insofar as it was relevant for the plaintiffs' standing to sue.

In the *Kivalina* claim, the plaintiffs were found to have no standing having failed to satisfactorily allege traceability – the causation element of the standing inquiry. The *Kivalina* court was sceptical on causation and in reference to the sequence of events allegedly leading to the plaintiffs' injury observed:

> [T]he harm from global warming involves a series of events disconnected from the discharge itself. In a global warming scenario, emitted greenhouse gases combine with other gases in the atmosphere which *in turn* results in the planet retaining heat, which *in turn* causes the ice caps to melt and the oceans to rise, which *in turn* causes the Arctic sea ice to melt, which *in turn* allegedly renders Kivalina vulnerable to erosion and deterioration resulting from winter storms.[488]

In *Comer*, standing was initially denied on, *inter alia*, causation grounds as the 'fairly traceable' element was not satisfied. Furthermore, the court noted:

> Without in any way expressing an opinion on the merits of the plaintiffs' claims against these defendants, I will observe that there exists a sharp difference of opinion in the scientific community concerning the causes of global warming, and

---

[486] *Comer v. Murphy Oil USA, Inc.*, 585 F.3d 855, C.A.5 (Miss.), 2009, p. 2.
[487] *Comer v. Murphy Oil USA, Inc.*, 585 F.3d 855, C.A.5 (Miss.), 2009, p. 5.
[488] *Native Village of Kivalina v. ExxonMobil Corp.*, 663 F.Supp.2d 863 (N.D.Cal. 2009), p. 8.

> I foresee daunting evidentiary problems for anyone who undertakes to prove, by a preponderance of the evidence, the degree to which global warming is caused by the emission of greenhouse gases; the degree to which the actions of any individual oil company, any individual chemical company, or the collective action of these corporations contribute, through the emission of greenhouse gases, to global warming; and the extent to which the emission of greenhouse gases by these defendants, through the phenomenon of global warming, intensified or otherwise affected the weather system that produced Hurricane Katrina.[489]

This statement indicates enormous evidentiary challenges to establishing a causal link between the defendant's GHG emissions, climate change, and the harm suffered by the plaintiff. In reversing the ruling of the district court, the Fifth Circuit found the *Comer* plaintiffs to have satisfied the causation requirement for Article III standing recognizing that the standard of proof for traceability was lower than that for causation on the merits: 'the Article III traceability requirement need not be as close as the proximate causation needed to succeed on the merits of a tort claim.'[490] Relying on *Massachusetts v. EPA*, the court noted that 'injuries may be fairly traceable to actions that *contribute* to, rather than solely or materially cause, greenhouse gas emissions and global warming.'[491]

In 2012, in its analysis of the 'fairly traceable' requirement for Article III standing, the district court was not convinced that 'the defendants' emissions caused or contributed to the specific damages [the plaintiffs] suffered during Hurricane Katrina.'[492] Yet, it engaged in an analysis of proximate cause '[a]ssuming for the sake of the argument only that the plaintiffs have alleged injuries that are fairly traceable to the defendants' conduct.' The court noted that proximate cause, defined as the 'cause which in natural and continuous sequence unbroken by any efficient intervening cause produces the injury and without which the result would not have occurred,' could not accommodate the plaintiffs' claims. It held that

> [t]he assertion that the defendants' emissions combined over a period of decades or centuries with other natural and man-made gases to cause or strengthen a hurricane and damage personal property is precisely the type of remote, improbable, and extraordinary occurrence that is excluded from liability.[493]

---

[489] *Comer v. Nationwide Mutual Insurance Co.*, WL 1066645 (S.D.Miss.) 23 February 2006), p. 4.
[490] *Comer v. Murphy Oil USA, Inc.*, 585 F.3d 855, C.A.5 (Miss.), 2009, p. 5, citations omitted.
[491] *Comer v. Murphy Oil USA, Inc.*, 585 F.3d 855, C.A.5 (Miss.), 2009, p. 6, emphasis in the original.
[492] *Comer v. Murphy Oil USA, Inc.*, Case 1:11-cv-00220-LG-RHW (S.D.Miss.), 20 March 2012.
[493] *Comer v. Murphy Oil USA, Inc.*, Case 1:11-cv-00220-LG-RHW (S.D.Miss.), 20 March 2012.

4.2.3.4 Attribution

As we saw earlier, the plaintiffs' claim in *California v. General Motors Corporation* did not make it past the early procedural phases and was dismissed on the grounds of non-justiciability of political questions. However, the court indicated that the causal link between the actions of the defendants and the climate change-induced harm caused to the plaintiffs would be difficult to prove and raised a related question of attribution:

> The Court is left without guidance in determining what is an unreasonable contribution to the sum of carbon dioxide in the Earth's atmosphere, or in determining *who should bear the costs* associated with the global climate change *that admittedly result from multiple sources around the globe.*[494]

The court noted the global nature of the climate change phenomenon and seemed to indicate that it would find it difficult to attribute to any particular defendant the harm caused by climate change. On the issue of damages, the court noted that the plaintiff's global warming nuisance tort claim sought to impose damages on an 'unprecedented' scale by 'grounding the claim in the pollution originating both within, and well beyond, the borders of the State of California.' The court added that it had no way of discerning the contributors to the alleged nuisance of global warming as there were 'multiple worldwide sources of atmospheric warming across myriad industries and multiple countries.'[495]

The *Kivalina* court made similar observations and admitted that since 'virtually everyone on Earth [was] responsible' for GHG emissions, the plaintiffs were in effect asking the court to make 'judgment that the two dozen Defendants [...] should be the only ones to bear the cost of contributing to global warming.'[496] The court further noted:

> Significantly, the source of the greenhouse gases are [*sic*] undifferentiated and cannot be traced to any particular source, let alone defendant, given that they rapidly mix in the atmosphere and inevitably merge with the accumulation of emissions in California and the rest of the world. [...] [I]t is not plausible to state which emissions – emitted by whom and at what time in the last several centuries and at what place in the world – caused Plaintiffs' alleged global warming related injuries.[497]

---

[494] *California v. General Motors Corp.,* 2007 WL 2726871 (N.D.Cal.), p. 15, emphasis added.

[495] *California v. General Motors Corp.,* 2007 WL 2726871 (N.D.Cal.), p. 15.

[496] *Native Village of Kivalina v. ExxonMobil Corp.,* 663 F.Supp.2d 863 (N.D.Cal. 2009), p. 10.

[497] *Native Village of Kivalina v. ExxonMobil Corp.,* 663 F.Supp.2d 863 (N.D.Cal. 2009), p. 13, references omitted.

The court in *Comer*, too, perceived as problematic the fact that the defendants' emissions were combined with other anthropogenic and natural GHGs released over a significant period of time.

4.2.3.5 Retroactivity

Related to the question of attribution is one other important issue that, in compensation claims, would likely be contentious on the merits – that of retroactivity. Retroactivity is likely to pose difficulties because such claims address actual harm, *i.e.* harm that has already occurred. It may be difficult to reconcile the fact that historic emissions from multiple sources have cumulatively led to climate change and its injurious consequences with attaching liability to a particular defendant. To hold a defendant responsible, it may be necessary to take into account past emissions released at the time when little was known about climate change and before the UNFCCC or IPCC came into existence. It is not clear how past emissions should be tackled as indeed it would be unreasonable to hold emitters liable for emissions discharged long before climate change scientists released their findings. In 1992, the UNFCCC – a framework agreement to reduce GHG emissions – was concluded and at least from that point on, climate change and its damaging consequences can be considered as clearly identified from the legal point of view. It has therefore been proposed to consider 1992 as a cut-off date for apportioning responsibility for climate change-related damage.[498] Another option would be to use 1994, the date of the UNFCCC's entry into force, as a cut-off date or the year it entered into force for the party concerned.

The phenomenon of cut-off dates is not new and the US tobacco cases may provide some helpful insights into the matter. In 2006, the Supreme Court of Florida decided *Engle v. Liggett Group*, a case arising from a smokers' class action lawsuit seeking damages against cigarette companies for smoking-related injuries. The court ruled that since the class could not be open-ended, a cut-off date for class membership was in order and ruled that 'the date of the trial court's November 21, 1996, order that recertified a narrower class [was] the appropriate cut-off date.'[499]

Regardless of whether or not a cut-off date for climate change liability is established, at least some degree of retroactive liability may be inevitable. It remains to be seen whether and how the courts will resolve those difficulties.

---

[498] D. Farber, 'Basic compensation for victims of climate change,' 38 *Envtl. L. Rep. News & Analysis* 10521, 2008, p. 10524.

[499] *Engle v. Liggett Group, Inc.,* 945 So.2d 1246 (Fla. 2006), p. 1255.

4.3 Concluding Remarks

The present chapter has demonstrated that legal challenges involved in the process of establishing climate change liability are experienced by all plaintiffs regardless of the type of relief they seek; however, the extent, and the exact content, of those challenges vary.

In claims related to procedural injury both, standing and causation, have been the subject of extensive judicial discussion. Yet, in claims related to procedural injury those issues have not been as problematic as in the other two categories of cases. For instance, the requirement of causation as an element of procedural standing in American courts has been relaxed and has not posed any significant challenges to plaintiffs. Having been extensively considered by Australian courts and, to some extent, by courts in New Zealand, causation – both general and specific – has been recognized in a number of procedural injury claims. This is indicative of a lower standard of proof involved in procedural cases due to the fact that no actual injury is at stake. It is significant that demonstrating that climate change must be taken into account by the relevant authority in approving a particular project or carrying out an EIA does not require the plaintiffs to meet the rigours of the *causa proxima* or 'but for' tests. This finding is particularly relevant for the analysis of interstate liability in the next chapter, particularly Micronesia's 2010 challenge of a power plant modernization project in the Czech Republic (see Section 5.2.3.3).

On the issue of attribution, the courts' pronouncements are scarce; yet, they seem to suggest that the multiplicity of emitters has little impact on procedural decisions concerning GHG emissions from a single source. It is noteworthy that, unlike claims for compensation or claims for injunctive and/or declaratory relief, procedural justice has not been hampered by the political question doctrine.

Actions for injunctive and/or declaratory relief have been somewhat less successful than claims related to procedural injury. The political question doctrine has presented some challenges; however, most difficulties have been associated with demonstrating standing and, particularly, causation as one of the requirements for standing. It is significant that the singular success of *Massachusetts v. EPA* appears to be rooted in the procedural character of the injury alleged by the plaintiffs. In practice, the US Supreme Court's ruling that the EPA did have the authority to regulate GHGs amounted to an injunction requiring the EPA to take regulatory action.

Since a number of injunctive relief claims have yet to be decided on the merits, it is unclear how the courts will approach certain substantive issues, particularly causation. The courts have indicated that on the merits, the standard of proof for demonstrating causation between the action complained of and the potential injury is necessarily higher than that for establishing standing.[500]

Claims for compensation constitute the least successful category of climate change-related claims. To a large extent, compensation claims have been hindered by the political question doctrine, which the courts have relied on to avoid making determinations of a political nature in accordance with the principles of the separation of powers. Standing and, particularly, causation as part of the standing inquiry have presented insurmountable challenges to the plaintiffs. It appears that courts have been more cautious in their approach to claims for compensation for actual damage as opposed to claims seeking injunctive relief to redress potential harm.

Should a compensation claim be decided on the merits, enormous evidentiary challenges to establishing a causal link between the defendant's GHG emissions and the actual harm suffered by the plaintiff are to be expected.[501] Attribution of harm to a particular defendant, too, is likely to present difficulties.[502] And last but not least, retroactivity issues would have to addressed, which may necessitate the introduction of a cut-off date.

---

[500] *E.g.*, see *Connecticut v. American Electric Power Co., Inc.*, 582 F.3d 309, C.A.2 (N.Y.), 2009, p. 61.
[501] *E.g.*, see *Comer v. Nationwide Mutual Insurance Co.*, WL 1066645 (S.D.Miss.) 23 February 2006, p. 4.
[502] *California v. General Motors Corp.*, 2007 WL 2726871 (N.D.Cal.), p. 15; *Native Village of Kivalina v. ExxonMobil Corp.*, 663 F.Supp.2d 863 (N.D.Cal. 2009), p. 13.

5 Establishing International Responsibility for Climate Change-Related Damage

5.1 Introduction

As has been made clear in Chapter 3, none of the existing approaches to state liability appear to offer any viable solutions with regard to climate change-related damage. Albeit the Kyoto Protocol is equipped with a compliance mechanism aimed at promoting, facilitating, and securing compliance of states parties with their obligations under the Protocol, it does not address the injurious consequences of climate change; proposals to include provisions to that effect have been rejected by industrialized nations. A proposal to assign to the KP Compliance Committee 'the power to require a state to pay for the restoration of damage to the environment' was not accepted.[503] Attempts to include the polluter-pays principle were likewise rejected.[504]

A comprehensive legally-binding global treaty remaining the preferred option for addressing anthropogenic climate change, the international negotiations have not advanced beyond the point of mandating the ADP to develop 'a protocol, another legal instrument or an agreed outcome with legal force under the Convention applicable to all Parties' by 2015 to take effect in 2020. With Canada, Japan, the Russian Federation, and New Zealand refusing to take on commitments under the KP second commitment period and developing countries having no emissions reduction or limitation commitments under the KP, the Protocol only covers approximately 15 per cent of the global emissions. Against this backdrop, no international claims have been brought to challenge heavily emitting states for their contribution to climate change. Attempts to hold such states to account have been unsuccessful. For example, in 2002, the island nation of Tuvalu announced its intention to launch lawsuits against the United States and Australia, neither of which had ratified the Kyoto Protocol at that time,[505] but subsequently abandoned the idea. In 2011, Palau and the Marshall Islands called upon the UN General Assembly to request the ICJ for an advisory opinion to clarify the obligations and responsibilities of states under international law for ensuring that GHG-emitting

---

[503] R. Lefeber, Climate Change and State Responsibility, *in* R. Rayfuse & S. Scott (Eds.), *International Law in the Era of Climate Change* (321-349), Cheltenham: Edward Elgar, 2012, p. 328.

[504] T.N. Slade, 'Climate change: the human rights implications for small island developing countries,' 37 *Environmental Policy and Law* 215, 2007, p. 218.

[505] 'Tuvalu threat,' ABC Local Radio, Australia, AM Archive, transcript from 4 March 2002, <www.abc.net.au/am/stories/s495507.htm> (last visited on 5 July 2012).

activities under their jurisdiction or control do not cause damage to other states.[506] Even if such an opinion is ultimately requested and delivered, the Court's advisory opinions are not legally binding, and for liability purposes, are nothing more than a reflection of the existing law.

Absent liability provisions in the Convention and the KP, the need to address loss and damage has now been recognized by the COP on several occasions and a work programme on loss and damage has been launched under the Convention.[507] Parties have also agreed to establish at COP19 institutional arrangements, such as an international mechanism, to address loss and damage in particularly vulnerable developing countries.[508] Yet, already now, it is submitted, climate change-related damage can be effectively addressed within the framework of the law of state responsibility.

Some academic writers have observed that the law of state responsibility has not 'played a large practical role in the environmental liability context' and that it is 'unlikely that a [s]tate responsibility approach could play a role in addressing global environmental problems' because 'the regular system of state responsibility is not particularly suitable for environmental protection.'[509] In the present author's view, this approach underrates the potential of the state responsibility framework to address effectively at least some environmental harms associated with, *inter alia*, the injurious effects of climate change. In fact, there has been a substantial increase in international environmental dispute settlement since the early arbitrations, such

---

[506] 'Palau seeks UN World Court opinion on damage caused by greenhouse gases,' UN News Centre, 22 September 2011, available from:
<www.un.org/apps/news/story.asp?NewsID=39710&Cr=pacific+island&Cr1=> (last visited on 5 July 2012).

[507] See *Bali Action Plan*, UNFCCC Decision 1/CP.13 (2007), para. 1(c)(iii) on consideration of 'disaster reduction strategies and means to address loss and damage associated with climate change impacts in developing countries that are particularly vulnerable to the adverse effects of climate change;' see also *Cancun Agreements: Outcome of the work of the* Ad Hoc *Working Group on Long-term Cooperative Action under the Convention*, UNFCCC Decision 1/CP.16 (2010), paras 25-26, 29 on recognizing 'the need to strengthen international cooperation and expertise in order to understand and reduce loss and damage associated with the adverse effects of climate change;' deciding 'to establish a work programme in order to consider [...] approaches to address loss and damage associated with climate change impacts in developing countries that are particularly vulnerable to the adverse effects of climate change;' and intending for the SBI to consider submissions to that effect 'with a view to making recommendations on loss and damage to the Conference of the Parties for its consideration at its eighteenth session.'

[508] *Approaches to address loss and damage associated with climate change impacts in developing countries that are particularly vulnerable to the adverse effects of climate change to enhance adaptive capacity*, FCCC/CP/2012/L.4/Rev.1 (UNFCCC decision number not available at the time of writing), 8 December 2012, para. 9.

[509] J. Brunnée, 'Of sense and sensibility: reflections on international liability regimes as tools for environmental protection,' 53 *Int'l & Comp. L.Q.* 351, 2004, pp. 351 and 353; J. Klabbers, Compliance Procedures, *in* D. Bodansky, J. Brunnée & E. Hey (Eds.), *The Oxford Handbook of International Environmental Law* (995-1009), New York: Oxford University, 2007, p. 1001.

as Pacific Fur Seal (1893), Trail Smelter (1941), and Lac Lanoux (1957),[510] and the law of state responsibility has played a progressively more significant role in the settlement of international environmental disputes.[511]

Dispute settlement under the law of state responsibility is without prejudice to MEAs' compliance procedures and mechanisms and, in principle, both processes could be engaged simultaneously or consecutively.[512] The potential role of the law of state responsibility *vis-à-vis* breaches of international obligations related to climate must be understood in the context of some important distinctions between compliance procedures and dispute settlement. First, compliance procedures extend beyond legal disputes whereas dispute settlement focuses on the interpretation and application of treaty provisions and obligations of states under customary international law. Under the Kyoto Protocol compliance mechanism, there need not be a legal dispute; it is a question of implementation submitted by an ERT, a party in respect of itself or a party in respect of another party that sets the compliance procedure into motion (see Chapter 2). In effectively enabling *actio popularis*, this method of triggering indicates that the aim of submitting a question of implementation to the KP Compliance Committee's EB is to protect common treaty interests and 'assist the defaulting state in returning to compliance,'[513] which is the second distinctive feature of compliance procedures. Third, unlike dispute settlement, compliance procedures are inherently non-adversarial;[514] they can be better described as dialogue based on peer review. Fourth, although the compliance procedures are compulsory, the results are not formally binding. It has been argued, however, that the non-binding character of the consequences applied by the Compliance Committee's EB does not affect their effectiveness. [515] This characteristic sets compliance procedures apart from such means of international dispute settlement as adjudication and arbitration, which require that states voluntarily submit to the jurisdiction of a competent court, whose decision will be legally binding on the parties to the dispute. Finally, compliance procedures are future-oriented and proactive. Once a compliance issue has been identified, the goal is to find its cause and formulate the response most appropriate for the state in question to remain in compliance or to return to a state of compliance as soon as

---

[510] B.H. Desai & B. Sidhu, 'International Environmental Dispute Settlement,' *Environmental Policy and Law*, 42/2, 2012, p. 106.

[511] Recent examples include: *Case Concerning the Gabčíkovo-Nagymaos Project (Hungary v. Slovakia)*, Judgment of 25 September 1997, 1997 ICJ Rep. 7; *Case Concerning Pulp Mills on the River Uruguay (Argentina v. Uruguay)*, Judgment of 20 April 2010, 2010 ICJ Rep. 14; *Responsibilities and Obligations of States with Respect to Activities in the Area*, Advisory Opinion of 1 February 2011, 2011 ITLOS Rep. 10.

[512] See 1997 KP, Art. 19 together with 1992 UNFCCC, Art. 14.

[513] M. Fitzmaurice & C. Redgwell, 'Environmental non-compliance procedures and international law,' 31 *Netherlands Yearbook of International Law* 35, 2000, p. 39.

[514] An exception is the EB of the KP Compliance Committee.

[515] Lefeber & Oberthür 2010, p. 151.

possible. Under the Kyoto Protocol, these responses range from facilitative to stronger measures, such as those applied by the EB (see Chapter 2). While dispute settlement based on the law of state responsibility is orientated towards the past in that it is aimed at redress, compliance regimes, on the contrary, focus mainly on future compliance. Although the procedure before the EB of the KP Compliance Committee may be seen 'among the non-compliance procedures, as the most similar to a judicial or arbitral one' (see Chapter 2) and although in practice, a determination of non-compliance by the EB amounts to a determination of breach of an international obligation, the ultimate goal remains that of incentivizing compliance in the future rather than determining the legal consequences of the breach alone.[516]

These distinctions suggest that there are several reasons for why dispute settlement under the law of state responsibility may be an attractive option to an injured state. First, since compliance procedures serve a common interest and not the interests of a single state, an injured state cannot obtain redress through a compliance procedure. However, this may be of less relevance in the context of climate change. As will be demonstrated below, most claims for reparation are unlikely to succeed and sustainable claims are likely to benefit mankind as a whole rather than a single state. Yet, a claim could still be brought in the interest of a particular group of states, such as SIDS. Second, adjudication or arbitration would secure a legally binding decision. Third, cessation and reparation of the injury caused by an internationally wrongful act in the form of a declaration of wrongfulness, could provide an effective remedy and has the potential to restore the international legal order.

Chapter 3 has drawn a distinction between state liability and state responsibility and outlined the existing approaches to interstate liability in international law, concluding that none of those approaches can be relied upon with respect to climate change-related damage. Now, Chapter 5 seeks to explain how the international law of state responsibility can provide an effective legal framework for dealing with the injurious consequences of climate change by way of imposing responsibility on the state for acting in breach of its international obligations. Since no interstate cases have been brought to date before an international court or tribunal, some core lessons will be drawn from climate change claims litigated in domestic courts (Chapter 4) insofar as parallels can be made.

---

[516] T. Treves, The Settlement of Disputes and Non-compliance Procedures, *in* T. Treves *et al.* (Eds.), *Non-Compliance Procedures and Mechanisms and the Effectiveness of International Environmental Agreements* (499-518), The Hague: TMC Asser Press, 2009, p. 510.

In order to enable the application of the rules of state responsibility in the climate change context, the relevant primary norms will be identified first. For this purpose, and pursuant to the distinction between developed and developing states reflected in the UNFCCC principle of common but differentiated responsibilities and respective capabilities,[517] four groups of states will be singled out: (1) industrialized countries participating in the Kyoto Protocol; (2) EITs; (3) industrialized countries not participating in the Kyoto Protocol; and (3) developing countries. Treaty-based obligations on climate change mitigation and climate change adaptation as well as the customary obligation to prevent significant transboundary harm will be considered for each group of states (Section 5.2). Second, once the primary obligations have been determined, the origins of international responsibility for each group of states will be dealt with (Section 5.3).[518] Third, the current chapter will address the content of state responsibility focusing on the legal consequences of the perpetrator state's failure to live up to its international obligations pertaining to mitigation of, and adaptation to, climate change as well as prevention of harm (Section 5.4). Fourth, questions of implementation of the international responsibility of the state in breach of its international obligations will be tackled (Section 5.5), which will be followed by some concluding remarks (Section 5.6).

5.2 Climate Change and Origins of State Responsibility: Primary Obligations of States

As has been demonstrated in Chapter 2, the international climate change regime contains no liability mechanism of its own. Despite their enforcement dimension, compliance procedures and mechanisms envisaged under the KP are geared towards the future and are primarily aimed at encouraging 'a non-complying State to return to compliance,'[519] and not to repair the injury caused by the wrongful conduct of a responsible state to the affected state. Albeit, in comparison with compliance regimes established under other MEAs, the KP compliance measures are the strongest, their nature is non-adversarial.[520] Significantly, the application of the KP compliance procedures is limited to states that are party to the Protocol, which leaves out developing states and developed states not party to the KP. Thus,

---

[517] See 1992 UNFCCC, preamble and Art. 3(1).

[518] A separate section is devoted to the origins of the international responsibility of EITs due to their special position.

[519] M. Fitzmaurice, 'The Kyoto Protocol compliance regime and treaty law,' 8 *Singapore Yearbook of International Law* 23, 2004, p. 25; G. Loibl, Compliance Procedures and Mechanisms, *in* M. Fitzmaurice, D.M. Ong & P. Merkouris (Eds.), *Research Handbook on International Environmental Law* (426-449), Cheltenham: Edward Elgar, 2010 , p. 429; see also Chapter 2.

[520] Loibl 2010, p. 435.

the only available framework for addressing internationally wrongful conduct, restoring the disrupted legal relation, and obtaining reparation for an international wrong is the law of state responsibility, which can provide interstate litigants with the necessary remedial mechanisms in the wake of an international breach. As Chapter 3 has shown, in the absence of primary rules governing liability for climate change-related damage, a state can only be held to account having committed an internationally wrongful act. In order to determine wrongfulness of a state's conduct, it must be established whether the act in question is attributable to that state and whether that act constitutes a breach of the relevant primary obligation. Therefore, the first step is to identify the primary obligations imposed on states by the global legal framework for international climate policy and customary international law. The next step would then be to analyse the legal consequences of their breach under the secondary norms of state responsibility.

5.2.1   International Obligations on Climate Change Mitigation

Today's international climate policy rests to a large extent on two pillars – mitigation and adaptation.[521] Having considered the IPCC Fourth Assessment Report, the UNFCCC Conference of the Parties at its thirteenth session decided 'to launch a comprehensive process to enable the full, effective and sustained implementation of the Convention through long-term cooperative action, now, up to and beyond 2012' by addressing, *inter alia* (1) enhanced action on mitigation of climate change; and (2) enhanced action on adaptation.[522]

An instrumental distinction to make in this context is that between industrialized and developing nations encapsulated in the UNFCCC's principle of common but differentiated responsibilities and respective capabilities. The global legal framework for climate policy makes this differentiation between developed and developing countries due to the fact that the largest share of historical and current global emissions of GHGs has originated in developed countries, that *per capita* emissions in developing states are still relatively low and that the developing countries' share of global emissions will continue to grow to meet their development needs.[523] The UNFCCC captures the essence of the principle of common but differentiated responsibilities as follows:

---

[521] 1992 UNFCCC, Art. 4; see also Chapter 2.
[522] *Bali Action Plan*, UNFCCC Decision 1/CP.13 (2007), especially para. 1(b) and 1(c); see also IPCC *AR4, Synthesis Report* (2007).
[523] See the 1992 UNFCCC, preamble.

> The Parties should protect the climate system for the benefit of present and future
> generations of humankind, on the basis of equity and in accordance with their
> common but differentiated responsibilities and respective capabilities. Accordingly,
> the developed country Parties should take the lead in combating climate change
> and the adverse effects thereof.[524]

In other words, it is accepted that the entire international community has a
responsibility to mitigate climate change; however, different countries bear various
degrees of responsibility, which is determined by their historic and contemporary
contributions as well as their implementation capacity.

The UNFCCC's ultimate objective is to stabilize GHG concentrations in the
atmosphere at a level that would prevent dangerous anthropogenic interference with
the climate system,[525] which is achieved through mitigation. According to the
Copenhagen Accord and Cancun Agreements, this objective must be interpreted to
mean that the increase in global average temperature is to be kept below 2°C above
preindustrial levels.[526] In order to stabilize GHG concentrations in the atmosphere,
the UNFCCC imposes upon industrialized states (UNFCCC Annex I parties) the
common quantified target of returning their anthropogenic GHG emissions to the
1990s levels by the end of the last decade of the twentieth century. Albeit achieved,
this target was not ambitious enough to stop climate change. Therefore, it was
agreed to supplement the Convention with a set of stricter quantified targets for
developed countries. Such new targets with a legally-binding effect found
manifestation in the 1997 Kyoto Protocol.[527] The revised target for industrialized
states provided for in the Kyoto Protocol aims at reducing their overall
anthropogenic emissions of GHGs by at least 5 per cent below 1990 levels in the
commitment period 2008-2012 and, in accordance with the KP amendment adopted
in Doha in 2012, by at least 18 per cent below 1990 levels in the commitment
period 2013-2020.[528] Additionally, the Kyoto Protocol assigns to developed

---

[524] 1992 UNFCCC, Art. 3(1).

[525] 1992 UNFCCC, Art. 2.

[526] *Copenhagen Accord*, UNFCCC Decision 2/CP.15 (2009), para. 1; *Cancun Agreements: Outcome of the
work of the* Ad Hoc *Working Group on Long-term Cooperative Action under the Convention*, UNFCCC
Decision 1/CP.16 (2010), para. 4.

[527] The need for revised targets was already recognized in 1995; see *Berlin Mandate: Review of the adequacy
of Article 4, paragraph 2(a) and (b), of the Convention, including proposals related to a protocol and
decisions on follow-up*, Decision 1/CP.1 (1995).

[528] 1997 KP, Art. 3(1) and Art. 3(1)*bis*: for KP amendment, see *Outcome of the work of the* Ad Hoc *Working
Group on Further Commitments for Annex I Parties under the Kyoto Protocol*, FCCC/KP/CMP/2012/L.9
(UNFCCC decision number not available at the time of writing), 8 December 2012.

countries (KP Annex B parties) individual quantified targets that must likewise be achieved by 2012 and 2020, respectively.[529]

However, not all UNFCCC Annex I parties are also included in Annex B of the KP.[530] On the basis of 1990 emissions, Annex B of the KP covered approximately 63.7 per cent of industrialised countries' emissions. This is due to the fact that the US, a major emitter among developed states, is not party to the Protocol and therefore, is not bound by its provisions. In 1990, the US was responsible for about 32 per cent of all industrialized countries' GHG emissions. By 2008, its share grew to approximately 39 per cent.[531]

In 2009, hopes were running high at the Copenhagen summit as COP15 negotiated towards an agreement that would provide for a comprehensive legally-binding emissions reduction strategy beyond 2012. At the end of the conference, which has widely been considered a failure, the COP took note of what is now known as 'the Copenhagen Accord' – a document without any binding force, which endorses the industrialized states' commitment to implementing individually or jointly the quantified economy-wide emissions targets for 2020. Set at 25-40 per cent below 1990 levels by 2020 in accordance with the IPCC recommendations, the aggregate target for industrialized countries has been repeatedly recognized by the CMP.[532] In Doha, it was agreed that industrialized countries taking on commitments under the second commitment period will revisit their commitments by 2014 and may increase their ambition in line with an aggregate reduction in GHG emissions of at least 25-40 per cent below 1990 levels by 2020.

Following Copenhagen, most industrialized and some developing countries subscribed to 2020 individual emissions reduction targets; however, those

---

[529] 1997 KP, Ann. B., as amended in 2012.

[530] In addition to the US, there are several other UNFCCC Annex I parties without KP Annex B targets. Due to its special circumstances recognized by the COP, Turkey is included in Annex I but not in Annex B. For Belarus, the amendment to Annex B concerning its emissions reduction target has not yet entered into force. Kazakhstan is party to the KP but does not have an emissions reduction target. Cyprus and Malta are EU member states but are not parties to the KP with a commitment inscribed in Annex B. On 15 December 2011, Canada, which in 1990 was responsible for about 3 per cent of developed states' emissions, withdrew from the Protocol. In addition to Canada, the Russian Federation, Japan, and New Zealand have not taken on commitments under the KP second commitment period.

[531] Calculated on the basis of *National Greenhouse Gas Inventory Data for the Period 1990-2008*, note by the secretariat, FCCC/SBI/2010/18 (4 November 2010), p. 18.

[532] See *Cancun Agreements: Outcome of the work of the* Ad Hoc *Working Group on Further Commitments for Annex I Parties under the Kyoto Protocol at its fifteenth session*, UNFCCC Decision 1/CMP.6 (2010), preambulary paragraphs; *Outcome of the work of the* Ad Hoc *Working Group on Further Commitments for Annex I Parties under the Kyoto Protocol at its seventh session*, UNFCCC Decision 1/CMP.7 (2011), preambulary paragraphs; and *Outcome of the work of the* Ad Hoc *Working Group on Further Commitments for Annex I Parties under the Kyoto Protocol*, FCCC/KP/CMP/2012/L.9 (UNFCCC decision number not available at the time of writing), 8 December 2012, para. 7.

voluntary targets are not legally binding and cannot entail responsibility in case of a
state's failure to comply with them.[533] Conversely, individual quantified emissions
limitation and reduction commitments taken on by industrialized countries for the
2013-2020 commitment period in accordance with the amendment to the KP
adopted in Doha in 2012 are binding, and non-attainment of those commitments
will amount to a breach of an international obligation (see below).

Although the implementation of quantified targets by developed countries is an
important element of mitigation of climate change, the UNFCCC's approach to
mitigation is broader and more inclusive. For example, it encompasses, *inter alia*,
conservation and forest management, development of energy efficiency, and
promotion of alternative energy sources. Taking into account their common but
differentiated responsibilities as well as specific development priorities, Article
4(1)(b) requires all states parties to formulate, implement, publish and regularly
update national and, where appropriate, regional programmes containing measures
to mitigate climate change.[534] However, the Convention imposes a further
obligation to adopt national policies and take corresponding measures on the
mitigation of climate change only on industrialized nations.[535] Furthermore,

> [t]he extent to which *developing* country Parties will effectively implement their
> commitments under the Convention will depend on the effective implementation
> by developed country Parties of their commitments under the Convention related
> to financial resources and transfer of technology and will take fully into account
> that economic and social development and poverty eradication are the first and
> overriding priorities of the *developing* country Parties.[536]

Thus, the developing states' obligation to formulate climate change mitigation
programmes is effectively limited and contingent on the financial resources and
transfer of technology by the developed states who have taken the lead in the
climate change mitigation process. Albeit the on-going economic and population
growth in the developing world is increasing the developing countries' share in the
global GHG emissions, their *per capita* emissions, are and, at least for some time,
will continue to be, lower than those in industrialized countries. For this reason,
developing states are indisposed to take on any legally binding mitigation

---

[533] For a list of industrialized countries that have subscribed to 2020 GHG emissions reduction targets see
<http://unfccc.int/home/items/5264.php> (last visited on 11 February 2011); a list of developing countries
that have undertaken similar commitments can be found at: <http://unfccc.int/home/items/5265.php>
(last visited on 11 February 2011); see also *Copenhagen Accord*, UNFCCC Decision 2/CP.15 (2009).
[534] 1992 UNFCCC, Art. 4(1)(b), see also 1997 KP, Art. 10(b).
[535] 1992 UNFCCC, Art. 4(2)(a), see also 1997 KP, Art. 2(1)(a).
[536] 1992 UNFCCC, Art. 4(7), emphasis added.

commitments. Nevertheless, following the conclusion of the Copenhagen Accord, a number of developing states have stated their intent of implementing nationally appropriate mitigation actions (NAMAs).[537] The Kyoto Protocol does not impose on developing countries any commitments to reach quantified emissions reduction or limitation targets. Yet, since some of them are already contributing to the global mitigation effort, the UNFCCC states parties have agreed that developing countries will continue to do so by taking NAMAs 'supported and enabled by technology, financing and capacity-building' provided by developed countries.[538] The objective of NAMAs is to achieve 'a deviation in emissions relative to 'business as usual' emissions in 2020.'[539] Thus, (1) the implementation of NAMAs by developing states is subject to developed countries' support, which the latter are under an obligation to give,[540] and (2) the contemplated emissions reduction or limitation by developing countries does not envision any legally binding quantified targets. It must be reiterated that developing states' commitments to NAMAs have no legally binding force; their voluntary character is repeatedly emphasized in the Cancun Agreements,[541] and failure to live up to them cannot entail state responsibility.

It is significant that under the UNFCCC and Kyoto Protocol, only industrialized countries are legally obligated to adopt national policies and take corresponding measures on the mitigation of climate change. Developing states are only bound by the general obligation to formulate national or regional programmes containing measures to mitigate climate change, whereas the obligation to implement such programmes is not binding and is further subject to the developed countries' provision of technological, financial, and capacity-building assistance. Therefore, under the UNFCCC and KP, only developed states can potentially be held responsible for failure to adopt national policies and take measures on climate change mitigation. Once attributed to a particular industrialized state, such an omission can potentially constitute a breach of an international obligation thereby giving rise to an internationally wrongful act, which in turn would entail international responsibility of that state (see Section 5.3.1). Whereas in such a case attributability would pose no difficulties, the question of breach of the obligation to adopt national policies and take corresponding measures on mitigation must be

---

[537] See <http://unfccc.int/home/items/5265.php> (last visited on 9 February 2011).

[538] *Cancun Agreements: Outcome of the work of the* Ad Hoc *Working Group on Long-term Cooperative Action under the Convention*, UNFCCC Decision 1/CP.16 (2010), paras 48-52.

[539] *Cancun Agreements: Outcome of the work of the* Ad Hoc *Working Group on Long-term Cooperative Action under the Convention*, UNFCCC Decision 1/CP.16 (2010), para. 48.

[540] *Cancun Agreements: Outcome of the work of the* Ad Hoc *Working Group on Long-term Cooperative Action under the Convention*, UNFCCC Decision 1/CP.16 (2010), para. 52.

[541] In particular, see *Cancun Agreements: Outcome of the work of the* Ad Hoc *Working Group on Long-term Cooperative Action under the Convention*, Decision 1/CP.16 (2010), paras 48, 50, and 54.

addressed with care. In order to establish whether this obligation has been breached, it is essential to determine its nature and exact content. In this respect, the UNFCCC and Kyoto Protocol can provide some guidance.

First, the obligation to adopt national policies and take corresponding mitigation measures appears to be both an obligation of conduct as well as an obligation of result. It is an obligation of result because it requires that industrialized nations put mitigation policies and measures in place, which is a result in itself. It is also an obligation of conduct in that it prescribes a course of conduct towards the attainment of its ultimate goal, that of climate change mitigation. Further, this obligation is binding on each industrialized state individually. While mitigation of climate change is its overall objective, compliance with this obligation would not be assessed in relation to the result of stopping, or significantly reducing, climate change. Rather, a state's conduct in adopting policies and taking mitigation measures would be measured against the standard of due diligence. However, reliance on due diligence in determining compliance with this obligation is controversial (see below).

Second, the obligation to adopt national policies and take corresponding measures on mitigation does not require industrialized states to *mitigate* climate change. Formulated in absolute terms, the obligation in question is only absolute in a procedural sense; it stops short of imposing an actual obligation to mitigate climate change. It has been argued that developed states may not even be required to exercise due diligence as long as they adopt some mitigation policies and measures.[542] As to the content of those policies and measures, the UNFCCC only provides that industrialized countries must limit their anthropogenic GHG emissions and protect and enhance their GHG sinks and reservoirs;[543] it offers no further explanation. The Kyoto Protocol requires that developed states implement and/or further elaborate policies and measures in accordance with their national circumstances and provides a non-exhaustive list of possible measures.[544] In other words, states enjoy complete discretion in the choice of domestic policies and measures on mitigation. Such policies and measures need to be assessed so that a determination could be made of whether or not a given state has complied with its obligations on mitigation. National communications submitted to the UNFCCC Conference of the Parties may be relevant in this regard; however, they cannot be relied upon for the purposes of determination of a breach of an international obligation. Instead, review by the Conference of the Parties is directed at the adequacy of the measures taken pursuant to the UNFCCC provisions.[545] Due to the

---

[542] Lefeber 2012, p. 331.
[543] 1992 UNFCCC, Art. 4(2)(a).
[544] 1997 KP, Art. 2(1)(a).
[545] 1992 UNFCCC, Arts. 4(2)(d), 7(2)(e), and 10(2)(b).

fact that the obligation to adopt policies and measures on mitigation does not require industrialized states to mitigate climate change (or even to act diligently to achieve this objective) and that the choice of such policies and measures is left to the states' discretion, it would be difficult to argue that a state has not complied with its UNFCCC and Kyoto Protocol obligation on mitigation as long as it adopts *some* policies and measures thereby conforming to the conduct prescribed.

Although the Kyoto Protocol does not obligate industrialized states to take specific mitigation measures but merely proffers a number of possible options, it imposes individual quantified emissions reduction or limitation targets that are legally binding and must be achieved by 2012 (the first commitment period) and by 2020 (the second commitment period). Non-attainment by a particular state of its individual target will lead to a breach of an international obligation. Industrialized countries are required to achieve their individual targets by domestic measures. The KP market-based supplementary mechanisms, *i.e.* emissions trading (Art. 17), joint implementation (Art. 6), and the CDM (Art. 12), provide an additional means for countries to meet their targets.[546]

Since the Kyoto Protocol affords industrialized countries complete freedom in the choice of measures towards the achievement of their emissions reduction or limitation targets, the obligation on individual quantified emissions reduction or limitation is an obligation of result, *i.e.* states' quantified targets must be achieved regardless of the exact means employed. The Kyoto Protocol's quantified approach also suggests that the industrialized states' obligations on mitigation are of an absolute character. Therefore, it would be relatively easy to determine whether a breach has occurred because if a state does not live up to its quantified emissions reduction or limitation target, it will have breached its international obligation under the Kyoto Protocol. However, whether or not a state has lived up to its emissions reduction or limitation targets under the first commitment period can be assessed only after 2012.[547] Taking into account the Kyoto Protocol reporting and review procedures, in practice such an assessment would not take place before 2015 (see Chapter 2).

To recap, only industrialized states are required to adopt mitigation policies and measures, however neither the UNFCCC nor the Kyoto Protocol specify the exact content of those measures. Therefore, in practice it may be difficult to prove that a developed state has breached its obligation to take mitigation measures as long as it has taken some steps towards that end. Developing states are not legally bound to

---

[546] See Chapter 2 for a detailed description of the KP market-based mechanisms.

[547] Accordingly, non-attainment of emissions limitation and reduction commitments under the KP second commitment period will not be assessed before 2020. Industrialized countries will also revisit their second commitment period commitments by 2014.

take mitigation measures but they may do so by adopting NAMAs, the preparation and implementation of which must be enabled by technology and finance provided by developed countries.[548] The Kyoto Protocol further imposes on industrialized countries an absolute obligation of result to achieve their respective quantified emissions reduction targets by 2012 under the first commitment period and by 2020 under the second commitment period. Cases of non-compliance will be addressed by the EB of the KP Compliance Committee.[549] An international claim raising questions of responsibility may be simultaneously or subsequently submitted to a dispute settlement body.[550] In the unlikely event of an injured state seeking reparation in the form of compensation, such a claim would have to be postponed until after the obligation to meet quantified emissions reduction or limitation targets is due, including the true-up period and the time necessary for the KP Compliance Committee to review country reports. However, if an injured state were to seek a declaration of wrongfulness, which, in combination with cessation of wrongful conduct, would effectively amount to an injunction, it would be sensible to bring a claim sooner than 2015. Yet, even a clear trajectory towards non-compliance, *e.g.* a continuous significant increase in an industrialized country's GHG emissions, cannot provide a legal basis for arguing that it has breached its obligations under the KP. Individual quantified emissions reduction or limitation targets are absolute obligations. Therefore, a possible breach cannot be assessed before the obligations are due, which in practice cannot happen before 2015, as explained above.

5.2.2 Obligations on Climate Change Adaptation

A certain degree of global warming is inevitable regardless of how successful climate change mitigation efforts are.[551] Adaptation addresses the adverse consequences of climate change that are taking place at the present time and aims at increasing mankind's resilience to future impacts. Under the UNFCCC, all states are required to implement measures to facilitate adequate adaptation to climate change and cooperate in preparing for adaptation to its impacts.[552] While successful mitigation is conditional on international cooperation, satisfactory adaptation can be achieved at the local level. However, as has been demonstrated in Chapter 1, climate change impacts are not uniform across the world and the worst effects will

---

[548] See *Cancun Agreements: Outcome of the work of the* Ad Hoc *Working Group on Long-term Cooperative Action under the Convention*, UNFCCC Decision 1/CP.16 (2010), para. 52.

[549] See 1992 UNFCCC, Art. 14(2) and 1997 KP, Art. 19.

[550] 1992 UNFCCC, Art. 14; on the relationship between the settlement of disputes and non-compliance procedures see Treves 2009.

[551] IPCC *AR4, Synthesis Report* (2007), p. 65.

[552] 1992 UNFCCC, Arts. 4(1)(b), 4(1)(e).

be suffered in developing countries despite their relatively small historical contribution thereto. Developing states also have the least capacity to adapt to the injurious consequences of climate change. Therefore, international cooperation is necessary in order to finance global adaptation efforts [553] and the central question is whether in undertaking those efforts industrialized countries are under any obligation to provide financial and technological assistance to developing states.

In principle, the UNFCCC and Kyoto Protocol require developed states to make available to developing countries financial resources for the development of adaptation policies and for the transfer of technology.[554] However, industrialized states appear to be under no obligation to finance the actual costs of adaptation measures in the developing world. They are merely required to endeavour to make technology available to developing states and to assist developing countries that are particularly vulnerable to the adverse effects of climate change in meeting costs of adaptation to those adverse effects.[555]

In 2007, the Bali Action Plan addressed enhanced action on adaptation including, *inter alia*, consideration of international cooperation.[556] In 2009, the non-legally binding Copenhagen Accord stressed 'the need to establish a comprehensive adaptation programme including international support.' [557] It reiterated the developed countries' intention to 'provide adequate, predictable and sustainable financial resources, technology and capacity-building to support the implementation of adaptation action in developing countries' and stated that 'new and additional, predictable and adequate funding as well as improved access shall be provided to developing countries [...] to enable and support enhanced action' on mitigation, adaptation, technology development and transfer as well as capacity-building.[558] The following year, COP16 adopted a decision on enhanced action and international cooperation on adaptation. For those purposes, the Cancun Adaptation Framework was established as well as a process to enable LDCs 'to formulate and implement national adaptation plans.'[559] Also, an Adaptation Committee was set up in order 'to promote the implementation of enhanced action on adaptation [...] under the Convention.' [560] Further, COP16 'requested' developed countries to

---

[553] Lefeber 2012, p. 326.

[554] 1992 UNFCCC, Art. 4(3) and 1997 KP, Art. 11(2).

[555] 1992 UNFCCC, Art. 4(3) & 4(4) and 1997 KP, Art. 11(2).

[556] *Bali Action Plan*, UNFCCC Decision 1/CP.13 (2007), para. 1(c).

[557] *Copenhagen Accord*, UNFCCC Decision 2/CP.15 (2009), para. 1.

[558] *Copenhagen Accord*, UNFCCC Decision 2/CP.15 (2009), paras 3, 8.

[559] *Cancun Agreements: Outcome of the work of the* Ad Hoc *Working Group on Long-term Cooperative Action under the Convention*, UNFCCC Decision 1/CP.16 (2010), paras 13-15.

[560] *Cancun Agreements: Outcome of the work of the* Ad Hoc *Working Group on Long-term Cooperative Action under the Convention*, UNFCCC Decision 1/CP.16 (2010), para. 20.

provide developing countries with 'long-term, scaled-up, predictable, new and additional finance, technology and capacity-building [...] to implement urgent, short-, medium- and long-term adaptation actions, plans, programmes and projects at the local, national, subregional and regional levels, in and across different economic and social sectors and ecosystems.'[561] At the same time, developed countries undertook a collective commitment of providing 'new and additional resources [...] approaching USD 30 billion for the period 2010-2012, *with a balanced allocation between adaptation and mitigation*' and with funding for adaptation being prioritized for the most vulnerable developing countries, such as LDCs, SIDS, and African states.[562] Since these short-term resources are intended for financing both adaptation and mitigation, the actual figure for adaptation efforts must be lower than the total. Also, this collective target is not translated into individual commitments. The non-binding language used in the Cancun Agreements (the COP 'requesting' industrialized states to make available finance and technology) and the *collective* pledge of funding point to the fact that developed states are under no individual legal obligation to provide financial resources to developing countries. Failure to follow up on the requests by the COP and on the collective commitments on adaptation financing set out in the Cancun Agreements cannot give rise to a breach of any individual international obligation by an industrialized state. Therefore, there is no legal basis for a finding of an internationally wrongful act and, consequently, state responsibility cannot be engaged.

According to Lefeber, a potential legal basis for holding industrialized countries liable for the financing of adaptation measures in developing states could be grounded in the polluter-pays principle, which is based on the assumption that the polluter is obligated to pay for the pollution it has caused.[563] This principle imposes the costs of the environmental harm on operators responsible for the pollution (see Chapter 3).[564] The polluter-pays principle is most easily applied in a geographical area subject to uniform environmental law, *e.g.* within one country or in the European Union;[565] however, it is unclear whether its cross-border

---

[561] *Cancun Agreements: Outcome of the work of the* Ad Hoc *Working Group on Long-term Cooperative Action under the Convention*, UNFCCC Decision 1/CP.16 (2010), para. 18.

[562] *Cancun Agreements: Outcome of the work of the* Ad Hoc *Working Group on Long-term Cooperative Action under the Convention*, UNFCCC Decision 1/CP.16 (2010), para. 95, emphasis added.

[563] R. Lefeber, *An Inconvenient Responsibility*, Utrecht: Eleven International Publishing, 2009, p. 13; Lefeber 2012, p. 326.

[564] 1992 Rio Declaration on Environment and Development, Principle 16. See also 1992 Convention for the Protection of the Marine Environment of the North-East Atlantic, UNTS, vol. 2354, p. 67 (2006), Art. 2(2)(b).

[565] Kiss & Shelton 2007a, pp. 96-97.

application extends to imposing financial liability on the state within whose territory the responsible polluter operates. In principle, it could be argued that since the state can and must control operators' activities within its territory, it should require operators to internalize transboundary costs of their activities because adaptation costs incurred in the developing world are aimed at the prevention of damage caused by them in the first place.[566] However, neither conventional law, nor case law supports such broad application of the polluter-pays principle. Therefore, industrialized countries' failure to finance adaptation measures in developing states cannot result in a breach of an international obligation and therefore cannot give rise to state responsibility.

As far as the issue of technology transfer is concerned, industrialized countries are not placed under an absolute obligation to transfer technologies to developing nations. Under the UNFCCC, developed countries are required to '*take all practicable steps* to promote, facilitate and finance, as appropriate, the transfer of, or access to, environmentally sound technologies and know-how' to developing countries to enable them to implement the provisions of the Convention.[567] In Cancun, industrialized states were encouraged to 'undertake domestic actions [...] to engage in bilateral and multilateral cooperative activities on technology development and transfer and to increase private and public research, development and demonstration in relation to technologies for mitigation and adaptation.'[568] Developed nations cannot be obligated to provide technologies to developing countries because private parties rather than states own the intellectual property rights on most technologies. In terms of domestic regulation, industrialized states may only be expected to invite private actors to make technologies available for the purposes of transferring them to developing nations; however, they cannot make this process compulsory. Several proposals on how to reconcile intellectual property rights with the need to accelerate development and transfer of technologies have been made by UNFCCC parties.[569] At present, failure to transfer technologies to developing states cannot result in a breach of an international obligation and thus, cannot give rise to the international responsibility of an industrialized state.

---

[566] Lefeber 2012, p. 327.

[567] 1992 UNFCCC, Art. 4(5), emphasis added.

[568] *Cancun Agreements: Outcome of the work of the* Ad Hoc *Working Group on Long-term Cooperative Action under the Convention*, UNFCCC Decision 1/CP.16 (2010), para. 116.

[569] *Ideas and proposals on paragraph 1 of the Bali Action Plan,* revised note by the Chair, FCCC/AWGLCA/2008/16/Rev.1 (15 January 2009), para. 129.

5.2.3   Obligation to Prevent Significant Transboundary Harm

It is submitted that states are under an obligation to take adequate measures to prevent transboundary damage from GHG emissions. This obligation is derived from the customary duty of states to ensure that activities within their jurisdiction or control do not cause transboundary damage.

The obligation not to cause transboundary damage originated in the Trail Smelter arbitration and is considered one of the foundational norms of contemporary international environmental law. [570] In ruling that 'under the principles of international law [...] no State has the right to use or permit the use of its territory in such a manner as to cause injury by fumes in or to the territory of another or the properties or persons therein, when the case is of serious consequence and the injury is established by clear and convincing evidence,'[571] the Trail Smelter arbitral tribunal gave expression to an important primary rule of international law.[572] The obligation to prevent transboundary damage stems from the concept of sovereign equality of states and prohibits states from causing significant harm in another's territory thereby protecting their respective national interests.

This principle is reflected in state practice and has been incorporated in a number of international environmental legal instruments. [573] For instance, it is reproduced as Principle 21 of the 1972 Stockholm Declaration on Human Environment and Principle 2 of the 1992 Rio Declaration on Environment and Development: 'States have [...] the responsibility to ensure that activities within their jurisdiction or control do not cause damage to the environment of other States or of areas beyond the limits of national jurisdiction.' Kiss & Shelton have further traced the principle through the 2002 World Summit on Sustainable Development, the UN Charter of Economic Rights and Duties of States, and the World Charter for Nature. They have further observed that the principle has been included in the 1982 Convention of the Law of the Sea,[574] the 1985 Association of South East Asian Nations (ASEAN) Convention on the Conservation of Nature and Natural Resources,[575] and the 1979 Geneva Convention on Long Range Transboundary Air

---

[570] Kiss & Shelton 2007a, p. 90.

[571] Trail Smelter case (United States v. Canada), 16 April 1938 and 11 March 1941, UNRIAA, vol. III 1905, at 1965.

[572] Drumbl 2006, p. 86.

[573] G. Handl, *Trail Smelter* in Contemporary International Environmental Law: Its Relevance in the Nuclear Energy Context, *in* R.M. Bratspies & R.A. Miller (Eds.), *Transboundary Harm in International Law* (125-139), New York: Cambridge University Press, 2006, p. 127.

[574] 1982 UNCLOS, Art. 194(2).

[575] 1985 ASEAN Agreement on the Conservation of Nature and Natural Resources, 15 *Environmental Policy and Law* 64 (1985), Art. 20.

Pollution.[576] It appears in the Convention on Biological Diversity[577] and the UNFCCC preamble.[578] The International Law Commission has relied on the no-harm principle in its recent work related to environmental law[579] and the ICJ has repeatedly endorsed it in its judgments. In its advisory opinion on the *Legality of the Threat or Use of Nuclear Weapons*, the ICJ declared that

> The existence of the general obligation of States to ensure that activities within their jurisdiction and control respect the environment of other States or of areas beyond national control is now part of the corpus of international law relating to the environment.[580]

The Court reiterated the significance of the obligation to prevent transboundary harm in the *Gabčíkovo-Nagymaros Project* judgment citing the above passage from its advisory opinion.[581] More recently, the obligation to prevent transboundary harm was reaffirmed by the ICJ in the *Pulp Mills* case and by the ITLOS Seabed Disputes Chamber in its advisory opinion on responsibilities and obligations of states sponsoring persons and entities with respect to activities in the area.[582] It can thus be concluded that the obligation to prevent transboundary harm is part of customary international law[583] and, as such, it is binding on all states and does not discriminate between developed and developing countries.

---

[576] 1979 Geneva Convention on Long-Range Transboundary Air Pollution, UNTS, vol. 1302, p. 217 (1992), preamble.

[577] 1993 CBD, Art. 3.

[578] Kiss & Shelton 2007a, p. 20.

[579] Draft Articles on the Law of the Non-Navigational Uses of International Watercourses, ILC Report on the work of its 46th session, A/49/10, YILC, vol. II, Part Two (1994), Art. 7 (Obligation not to cause significant harm); 2001 Draft Articles on Prevention of Transboundary Harm from Hazardous Activities, Art. 3 (Prevention); Draft Articles on the Law of Transboundary Aquifers, ILC Report on the work of its 60th session, A/63/10, forthcoming in YILC (2008), Art. 6 (Obligation not to cause significant harm); see also Chapter 5.

[580] *Legality of the Threat or Use of Nuclear Weapons,* Advisory Opinion of 8 July 1996, 1996 ICJ Rep. 226, at 241, para. 29.

[581] *Case Concerning the Gabčíkovo-Nagymaos Project (Hungary v. Slovakia),* Judgment of 25 September 1997, 1997 ICJ Rep. 7, at 41, para. 53.

[582] *Case Concerning Pulp Mills on the River Uruguay (Argentina v. Uruguay),* Judgment of 20 April 2010, 2010 ICJ Rep. 14, at 79, para. 197; *Responsibilities and Obligations of States with Respect to Activities in the Area,* Advisory Opinion of 1 February 2011, 2011 ITLOS Rep. 10, paras 110-120.

[583] Fitzmaurice 2007, p. 1013; Handl 2006, pp. 126-129; N. de Sadeleer, The Principles of Prevention and Precaution in International Law: Two Heads of the Same Coin? *In* M., Fitzmaurice, D.M. Ong & P. Merkouris (Eds.), *Research Handbook on International Environmental Law* (182-199), Cheltenham: Edward Elgar, 2010, p. 182.

5.2.3.1 Obligation to Prevent Transboundary Harm as a Duty of Due Diligence

The obligation to prevent transboundary harm can also be extended to damage caused by the emission of GHGs[584] and its potential to require states to mitigate climate change is of particular consequence. The states' duty to ensure that activities within their jurisdiction or control do not cause transboundary harm is a due diligence obligation and does not impose an absolute duty to prevent harm.[585] Whereas prevention of transboundary damage is the objective of this obligation, it is not an obligation to achieve, in each and every case, the *result* of harm prevention. Rather, as a due diligence obligation of conduct, it is 'an obligation to deploy adequate means, to exercise best possible efforts, to do the utmost, to obtain this result.'[586]

States are required to exercise due diligence to achieve the objective of prevention of damage resulting from GHG emissions associated with activities within their jurisdiction or control. Compliance with the obligation to prevent transboundary harm necessitates the adoption, implementation, and enforcement of certain policies and measures. In its judgment in the *Pulp Mills on the River Uruguay* case, the ICJ has illustrated the meaning of the obligation to act with due diligence as follows:

> It is an obligation which entails not only the adoption of appropriate rules and measures, but also a certain level of vigilance in their enforcement and the exercise of administrative control applicable to public and private operators, such as the monitoring of activities undertaken by such operators [...].[587]

---

[584] C. Schwarte & R. Byrne, International Climate Change Litigation and the Negotiation Process, Foundation for International Environmental Law and Development working paper, 2010, available from: <www.field.org.uk/files/FIELD_cclit_long_Oct.pdf> (last visited on 1 March 2011), pp. 6-7; R. Tol & R. Verheyen, Liability and Compensation for Climate Change Damages – a Legal and Economic Assessment, working paper FNU-9, Research unit Sustainability and Global Change, Hamburg University, 2001, available from: <www.fnu.zmaw.de/fileadmin/fnu-files/publication/working-papers/adapcap.pdf> (last visited on 1 March 2011), p. 12; C. Voigt, 'State responsibility for climate change damages,' 77 *Nordic Journal of International Law* 1, 2008, pp. 7-9.

[585] *Case Concerning Pulp Mills on the River Uruguay (Argentina v. Uruguay)*, Judgment of 20 April 2010, 2010 ICJ Rep. 14, at 79, para. 197; *Responsibilities and Obligations of States with Respect to Activities in the Area*, Advisory Opinion of 1 February 2011, 2011 ITLOS Rep. 10, paras 110-111; Kiss & Shelton 2007a, p. 91.

[586] *Responsibilities and Obligations of States with Respect to Activities in the Area*, Advisory Opinion of 1 February 2011, 2011 ITLOS Rep. 10, para. 110.

[587] *Case Concerning Pulp Mills on the River Uruguay (Argentina v. Uruguay)*, Judgment of 20 April 2010, 2010 ICJ Rep. 14, at 79, para. 197; see also *Responsibilities and Obligations of States with Respect to Activities in the Area*, Advisory Opinion of 1 February 2011, 2011 ITLOS Rep. 10, paras 110-116.

Thus, in fulfilling their due diligence obligation to prevent transboundary harm, states are under an obligation to take regulatory and enforcement measures towards compliance. With respect to climate change, states are obligated to adopt policies and measures preventing, limiting or reducing GHG emissions in order to comply with their general due diligence obligation to prevent transboundary damage resulting from those emissions.

As discussed in Section 5.2.1 above, the UNFCCC obligation to adopt policies and take corresponding measures on climate change mitigation (binding only on industrialized states) does not actually require that climate change is successfully mitigated. It leaves developed states with a great deal of leeway in the choice of measures to take. States are nonetheless bound by the customary obligation to prevent transboundary damage and are required to exercise due diligence to that end. As has been mentioned earlier, the obligation to prevent significant transboundary harm is binding on developed and developing states alike. In exercising due diligence, states are required to regulate public and private conduct in areas subject to their jurisdiction or control that may cause transboundary environmental harm. In the international climate policy context, this must be done through mitigation action that may or may not lead to significant transboundary harm prevention; however, states must show due diligence in their endeavours to avoid it.

Now that it has been established that states are required to exercise due diligence in fulfilling their obligation to prevent transboundary harm resulting from GHG emissions, it is important to estimate the degree of due diligence associated with this obligation. The due diligence standard can be described from an objective and subjective point of view. Objectively, the degree of due diligence exercised by a state in order to prevent transboundary harm must not be 'significantly less' than the degree of diligence other states may be expected to exercise.[588] Therefore, in principle, all states are required to be compatibly diligent in acting towards transboundary harm prevention. Subjectively, however, the degree of diligence due in a particular situation must depend on the circumstances of the case, available means, and the nature of protected interests.[589] For example, the ITLOS recently recognized the possibility of different treatment of developed and developing states and that 'the requirements for complying with the obligation to apply the precautionary approach may be stricter for the developed than for the developing sponsoring States;' however, it also observed that what counts in a particular situation is 'the level of scientific knowledge and technical capability available to a

---

[588] Lefeber 2012, p. 335.
[589] British Claims in the Spanish Zone of Morocco (Spain v. United Kingdom), 1 May 1925, UNRIAA, Vol. II, p. 615 (2006), at 644, para. 4.

given State in the relevant scientific and technical fields.'[590] The subjective
approach to GHG emission regulation in the international climate policy context is
reflected in the UNFCCC and KP provisions on mitigation that distinguish between
developing and industrialized states and rely on the principle of common but
differentiated responsibilities and respective capabilities.[591] Conversely, the special
legal regime under the UNCLOS makes no such distinction and hence does not
differentiate between the degrees of due diligence on the basis of levels of
development. With respect to the international climate regime, it may be argued
that the degree of due diligence to be deployed by industrialized states is not the
same as can be expected of developing countries. This is not to suggest that
developing states are freed from the obligation to act diligently for the purposes of
transboundary harm prevention but the subjective approach does seem to imply that
the degree of due diligence countries must exercise corresponds to 'an objective
international standard for states with an equivalent level of prosperity.'[592]

It is also important to keep in mind that due diligence is a variable concept. It
'may change over time as measures considered sufficiently diligent at a certain
moment may become not diligent enough in light [...] of new scientific or
technological knowledge.'[593] Thus, with the development of climate change science,
the degree of due diligence required to meet the obligation to prevent
transboundary damage from GHG emissions may increase for industrialized and
developing states, respectively. For example, if in 1990 scientists cautioned that
there merely was 'concern that human activities may be inadvertently changing the
climate of the globe through the enhanced greenhouse effect,'[594] in 2007 they stated
with 'very high confidence that the net effect of human activities since 1750 has
been one of warming,' that most of the observed increase in global temperatures
since the mid-20th century was 'very likely' due to the increase in anthropogenic
GHG concentrations, and that discernible human influences also extended to other
aspects of climate (*e.g.* sea level rise, changes in weather patterns, heat waves, and
heavy precipitation events).[595] These scientific developments have inspired the
advancement of the international climate policy and continue to inform the ongoing
negotiations. The UNFCCC common quantified target of returning anthropogenic
GHG emissions to the 1990s levels by the end of the last decade of the twentieth
century, albeit achieved, turned out to be insufficient to prevent dangerous

---

[590] *Responsibilities and Obligations of States with Respect to Activities in the Area*, Advisory Opinion of
1 February 2011, 2011 ITLOS Rep. 10, paras 160-162.
[591] 1992 UNFCCC, Art. 3(1).
[592] Lefeber 2012, p. 335.
[593] *Responsibilities and Obligations of States with Respect to Activities in the Area*, Advisory Opinion of
1 February 2011, 2011 ITLOS Rep. 10, para. 117.
[594] IPCC *FAR, Climate Change: The IPCC Scientific Assessment*, Policymakers Summary (1990), p. xiii.
[595] IPCC *AR4, Synthesis Report*, Summary for Policymakers (2007), pp. 5-6.

anthropogenic interference with the climate system. Thus, what would have been considered diligent in the 1990s can no longer be regarded as diligent today as is evidenced by the new scientific data reflected in the IPCC reports, KP commitments undertaken by industrialized countries for the two commitment periods, and the current negotiations towards a new agreement.

Furthermore, the IPCC AR4 has shown that despite the current mitigation policies, global GHG emissions are growing and will continue to grow over the next few decades.[596] This means that the industrialized countries' commitments under the KP and Cancun Agreements and the developing countries' NAMAs under the Cancun Agreements are no longer sufficient to prevent significant transboundary harm that will result from the changing climate unless more ambitious measures are taken. Therefore, it is submitted that the due diligence standard for significant harm prevention must be higher than what can be achieved under the current international climate policy.

5.2.3.2 Scope of the Obligation to Prevent Transboundary Harm

Bearing in mind the climate change mitigation potential of the obligation to prevent transboundary harm, its scope calls for further elaboration. It must first be noted that the obligation to prevent transboundary damage extends to all activities within a state's jurisdiction or control and is not limited to hazardous activities which are regulated by a special set of rules.[597] Hazardous activities are activities that have a high probability of causing significant transboundary harm or a low probability of causing disastrous transboundary harm.[598] The emission of GHGs from a particular source is not hazardous *per se* as global warming is caused by an accumulation of GHGs from multiple activities as well as gradual degradation of GHG sinks worldwide. Therefore, states must address the potential dangerous effects of GHG emissions by regulating activities within their jurisdiction or control. By taking such regulatory measures, a state would exercise due diligence required of it by the obligation to prevent significant transboundary harm. Such measures derive from procedural and substantive duties associated with the obligation to prevent transboundary harm and are analysed in Section 5.2.3.3.

---

[596] IPCC *AR4, Synthesis Report*, Summary for Policymakers (2007), p. 72; see also Chapter 1.

[597] 2006 Principles on the Allocation of Loss in the Case of Transboundary Harm Arising out of Hazardous Activities; see Chapter 4.

[598] 2006 Principles on the Allocation of Loss in the Case of Transboundary Harm Arising out of Hazardous Activities, ILC Report on the work of its 58th session, A/61/10, forthcoming in YILC, pp. 116-117.

Second, the scope of the obligation not to cause transboundary damage is such that it does not only apply to activities within a state's territory but to any areas within its jurisdiction or control.[599] That includes GHG emissions by marine vessels and aircraft flying its flag. The Kyoto Protocol provides for the reduction of GHG emissions in the aviation and marine transport sectors to be carried out through the ICAO and IMO, respectively.[600] Both organizations have since taken steps towards improving fuel efficiency and sector-specific GHG emissions reduction. In its consolidated statement of policies and practices related to environmental protection, the ICAO, without attributing specific obligations to individual states and without prejudging the outcome of negotiations under the UNFCCC and KP, has undertaken to achieve a global annual average fuel efficiency improvement of 2 per cent until 2020 and an aspirational global fuel efficiency improvement rate of 2 per cent annually in the 2021-2050 period.[601] The IMO's Marine Environment Protection Committee (MEPC) has adopted a package of interim guidelines on specific technical and operational reduction measures to improve energy efficiency of ships.[602] Both organizations are also working on the development of market-based mechanisms for industry-specific GHG emissions reduction.[603]

5.2.3.3 Duties Stemming from the Obligation to Prevent Transboundary Harm

Structurally, the obligation to prevent transboundary damage is a composite one as compliance with it involves a variety of procedural and substantive duties.[604] In

---

[599] *Case Concerning the Gabčikovo-Nagymaos Project (Hungary v. Slovakia)*, Judgment of 25 September 1997, 1997 ICJ Rep. 7, at 41, para. 53.

[600] 1997 KP, Art. 2(2).

[601] *Consolidated statement of continuing ICAO policies and practices related to environmental protection – General provisions, noise and local air quality*, ICAO Assembly Resolution A37-18 (2010), para. 1(c); *Consolidated statement of continuing ICAO policies and practices related to environmental protection – Climate change*, ICAO Assembly Resolution A37-19 (2010), preambulary paragraphs and paras 4-5.

[602] IMO Marine Environment Protection Committee circulars: *Interim guidelines on the method of calculation of the energy efficiency design index for new ships*, IMO Marine Environment Protection Committee, MEPC.1/Circ.681 (17 August 2009); *Interim guidelines for voluntary verification of the energy efficiency design index*, IMO Marine Environment Protection Committee, MEPC.1/Circ.682 (17 August 2009); *Guidance for the development of a ship energy efficiency management plan (SEEMP)*, IMO Marine Environment Protection Committee, MEPC.1/Circ.683 (17 August 2009); and *Guidelines for voluntary use of the ship energy efficiency operational indicator (EEOI)*, IMO Marine Environment Protection Committee, MEPC.1/Circ.684 (17 August 2009).

[603] See *Consolidated statement of continuing ICAO policies and practices related to environmental protection – Climate change*, ICAO Assembly Resolution A37-19 (2010), paras 13-18 and Ann.; for information on IMO initiatives see <www.imo.org/OurWork/Environment/PollutionPrevention/AirPollution/Pages/GHG-Emissions.aspx> (last visited on 12 September 2011).

[604] Kiss & Shelton 2007a, p. 91.

acting diligently, a state must regulate activities within its jurisdiction or control through general measures or measures pertaining to the authorization of a particular activity. Among procedural duties, Lefeber lists assessment of the transboundary impact of a particular measure;[605] notification of potentially affected states; exchange of information with the states involved; consultations and negotiations with those states; and the monitoring of transboundary environmental impacts throughout the implementation stage of the relevant measure.[606] Given the fact that GHG emissions from a particular source contribute to global climate change that may potentially affect the entire international community, including the source state, states that are likely to be affected have the right to demand that other states comply with their implementation duties. Recently, the Federated States of Micronesia made use of this right.

In a letter to the Ministry of the Environment of the Czech Republic dated 4 January 2010, the government of the Federated States of Micronesia expressed its concern over the plan for the modernization of the lignite-fired power plant Prunéřov II.[607] An Environmental Impact Assessment (EIA) of the proposed modernization plan had been made but it did not take into account the transboundary impact of the project. The Federated States of Micronesia requested that the Czech Ministry of the Environment issue a negative final statement on the EIA of the plan. Micronesia stated that the Prunéřov modernization plan was in substantial conflict with the applicable EU and Czech law. In Micronesia's view, Prunéřov, being one of the largest single sources of GHG emissions in the world and the largest one in the Czech Republic, contributed significantly to climate change and its serious environmental impacts could affect the territory of the Micronesian state. Micronesia underscored that it was particularly vulnerable to the dangerous impacts of climate change, particularly flooding, and that Prunéřov could dangerously affect its environment by contributing to the accelerated sea-level rise. On 1 February 2010, the Czech Minister of the Environment decided to submit the project for independent international assessment without any explicit mention of Micronesia's request.[608] Yet, the subsequent EIA of the project that the Minister accepted on 29 April 2010 did not take into account its transboundary

---

[605] See Micronesia's challenge to the Prunéřov plant modernization plan *infra*.

[606] Lefeber 2012, p. 335.

[607] Viewpoint of the Federated States of Micronesia on the complex renovation of Prunéřov II power plant 3x250 MWe plan, letter of 4 January 2010 from the Director of the Office of Environment and Emergency Management of the Federated States of Micronesia to the Ministry of the Environment of the Czech Republic, available from: <http://aktualne.centrum.cz/domaci/fotogalerie/2010/01/06/dokument-dopis-mikronesie-k-planu-na-rekonstrukci-/foto/287153/?cid=657469> (last visited on 14 March 2011).

[608] Ministry of the Environment of the Czech Republic press release of 1 February 2010, available from: <www.mzp.cz/en/news_100126> (last visited on 9 March 2011).

impacts.[609] Incidentally, this happened shortly after the ICJ ruling in the *Pulp Mills on the River Uruguay* case, in which the Court referred to

> [...] a practice, which in recent years has gained so much acceptance among States that it may now be considered a requirement under general international law to undertake an environmental impact assessment where there is a risk that the proposed industrial activity may have a significant adverse impact in a transboundary context, in particular, on a shared resource. Moreover, due diligence, and the duty of vigilance and prevention which it implies, would not be considered to have been exercised, if a party planning works liable to affect the régime of the river or the quality of its waters did not undertake an environmental impact assessment on the potential effects of such works.[610]

Although the *Pulp Mills on the River Uruguay* judgment concerned a specific situation submitted to the Court, reference to the customary status of the requirement to make an EIA in a transboundary context is significant; in its absence, the duty of due diligence cannot be considered fulfilled. The following year, the ITLOS also confirmed that the obligation to conduct an EIA was a general obligation under customary international law.[611]

Substantive duties stemming from the composite obligation to prevent transboundary damage are largely determined by the source of damage.[612] In order to prevent significant transboundary harm associated with climate change, states are required to exercise due diligence by way of taking mitigation action in the form of aggregate emissions reduction. Limiting GHG emissions may involve the use of a cap-and-trade mechanism, carbon capture and storage or the use of the best available technology (BAT) standard by single major emissions sources.

The latter standard is of relevance to Micronesia's challenge of the Prunéřov modernization plan. Whereas Micronesia did not contest the modernization project *per se*, it demanded that the BAT standard be used for the Prunéřov modernization as set out in both the EU and Czech legislation on net energy efficiency of new power plants, which requires a minimum of 42 per cent net energy efficiency for a new power plant while the EIA proposed only 38 per cent. Micronesia expressed its concern over this discrepancy and questioned whether the renovation of Prunéřov was not in fact a new construction, which would be subject to higher limits of

---

[609] Ministry of the Environment of the Czech Republic press release of 30 April 2010, available from: <www.mzp.cz/en/news_100430_statement_Prunerov> (last visited on 9 March 2011).

[610] *Case Concerning Pulp Mills on the River Uruguay (Argentina v. Uruguay)*, Judgment of 20 April 2010, 2010 ICJ Rep. 14, at 82, para. 204.

[611] *Responsibilities and Obligations of States with Respect to Activities in the Area*, Advisory Opinion of 1 February 2011, 2011 ITLOS Rep. 10, para. 145.

[612] Lefeber 2012, p. 336.

efficiency than reconstruction projects.[613] Although, as has been mentioned earlier, the EIA accepted by the Czech Minister of the Environment included no transboundary impact assessment, its approval was subject to a condition aimed to compensate the use of 'technology with lower effectiveness than that of the best available techniques.'[614] In fact, the compensatory measures proposed and eventually accepted 'will not only attain, but even exceed by 84 %, the required savings of greenhouse gases.'[615] It must be clarified that while the use of technology with lower effectiveness than the BAT in combination with certain compensatory measures may have been justified by the Minister of the Environment with respect to the modernization of an existing plant, the Czech Environmental Ministry does not appear to contest the fact that the approval of new plants construction proposals must be determined by the BAT standard. The Minister's decision to subject the Prunéřov modernization project to the requirement to take compensatory measures to counterbalance the use of inferior technology seems to signify the implicit intention of the Czech Republic to exercise due diligence in abiding by the duty to prevent significant transboundary harm.

5.2.3.4 Significant Harm Requirement

The obligation to prevent transboundary harm is qualified by the degree of actual or potential damage: it only requires states to prevent transboundary harm that is significant. To quote the ILC, '[t]he ecological unity of the planet does not correspond to political boundaries [and i]n carrying out lawful activities within their own territories, States have impacts on each other. These mutual impacts, so long as they have not reached the level of "significant," are considered tolerable.'[616] Thus, some level of transboundary harm, as long as it is not significant, is acceptable.

Chapter 1 has described the observed and projected changes in climate and their impacts. It has been concluded that many of those impacts are having, and will

---

[613] Viewpoint of the Federated States of Micronesia on the complex renovation of Prunéřov II power plant 3x250 MWe plan, letter of 4 January 2010 from the Director of the Office of Environment and Emergency Management of the Federated States of Micronesia to the Ministry of the Environment of the Czech Republic referencing Directive 2008/1/EC of the European Parliament and of the Council of 15 January 2008 concerning integrated pollution prevention and control, OJ 2008 L 24/8, and Reference Document on Best Available Techniques for Large Combustion Plants.

[614] Ministry of the Environment of the Czech Republic press release of 20 October 2010, available from: <www.mzp.cz/en/news_101020_Prunerov> (last visited on 14 March 2011).

[615] Ministry of the Environment of the Czech Republic, press release of 20 October 2010.

[616] 2006 Principles on the Allocation of Loss in the Case of Transboundary Harm Arising out of Hazardous Activities, ILC Report on the work of its 58th session, A/61/10, forthcoming in YILC, commentary to Principle 2, para. 2.

continue to have, devastating effects, particularly in the developing world. The magnitude of those injurious impacts will vary and may be significantly reduced depending on the effectiveness of mitigation and adaptation measures and the rate of the global average temperature change; however, the injurious consequences of climate change cannot be eliminated entirely.[617] The types of harm associated with those impacts include loss of land and livelihoods due to the rising sea levels and consequent displacement of persons and/or mass migration, damage from extreme weather events and changes in water availability patterns (*e.g.* floods, droughts, storms, hurricanes), negative impacts on human health and food sources associated with the global temperature rise (*e.g.* malnutrition and food shortages due to decreases in crop productivity, increased burden from infectious diseases), and damage to ecosystems.[618]

It is submitted that climate change damage – be it damage to the environment or human health, loss of territory or damage to crops, loss of biodiversity or wide-ranging damage to economic interests – falls within the protective scope of the obligation to prevent significant transboundary harm. The actual occurrence of damage is not required as long as it can be shown that a state's acts or omissions have the potential of resulting in such damage[619] and, according to the IPCC Fourth Assessment Report, climate change damaging impacts are imminent and cannot be underestimated.[620]

The significance requirement for transboundary damage is well established in international environmental law.[621] However, there are no strict criteria for determining the exact threshold of significant harm. The concept of a threshold of damage was introduced in the Trail Smelter case, which referred to injury of 'serious consequence.'[622] Following that decision, sensitivity of the international community to transboundary environmental damage has undergone considerable changes. The ILC has defined significant damage as something more than detectable or appreciable, but not necessarily serious or substantial. According to that understanding, the harm must lead to a real detrimental effect on human health, industry, property, environment or agriculture in other States and such detrimental effects must be susceptible of being measured by factual and objective standards.[623]

---

[617] IPCC *AR4, Synthesis Report* (2007), p. 65.

[618] IPCC *AR4, Synthesis Report*, Summary for Policymakers (2007), p. 10.

[619] Lefeber 2012, p. 338.

[620] See Chapter 1.

[621] See, *e.g.* Handl 2006; Lefeber 2012.

[622] Trail Smelter case (United States *v.* Canada), 16 April 1938 and 11 March 1941, UNRIAA, vol. III 1905, at 1965.

[623] 2001 Articles on Responsibility of States for Internationally Wrongful Acts, ILC Report on the work of its 53rd session, A/56/10, YILC, vol. II, Part Two, commentary to Art. 2, paras 4 and 5; 2006 Principles on the

This is a reflection of the *de minimis* approach to the determination of the threshold. Another approach involves case-by-case assessment of all economic, social, and environmental circumstances and proposes taking into account all interests involved by 'balancing the socio-economic utility of an activity against its detrimental environmental effects.'[624] Such a threshold requirement balances the right of states to develop against the obligation to prevent transboundary damage.[625]

With respect to climate change damage, the accepted increase in global temperature of not more than 2°C above preindustrial levels can be helpful in determining the significance-of-harm threshold. Throughout the UNFCCC negotiating process, states have acknowledged the fact that a lesser global temperature rise will still result in a certain degree of climate change-associated harm. However, the international community has agreed to tolerate the extent of damage associated with the global temperature increase of less than 2°C. Anything above this level would amount to *significant* harm. Given the current mitigation effort, the global temperature rise will reach, and possibly exceed, the 2 degree threshold by the end of this century.[626] Thus, the potential additional harm resulting from climate change must be considered significant.

The ILC has observed that while a certain 'deprivation at a particular time might not be considered 'significant' because scientific knowledge or human appreciation at that specific time might have considered such deprivation tolerable,' that view may later change and 'the same deprivation might then be considered 'significant damage.''[627] In this regard, it is of consequence that the UNFCCC Conference of the Parties has recognized the need to revise the long-term goal for the acceptable global temperature increase, including in relation to the proposed target of 1.5°C.[628] Whether or not this revision is put into practice, states must at the very least take mitigation action in order to prevent harm associated with the global temperature rise of more than 2°C.

The obligation to prevent transboundary harm could be relied upon in contesting the respondent state's failure to take adequate mitigation action in

---

Allocation of Loss in the Case of Transboundary Harm Arising out of Hazardous Activities, ILC Report on the work of its 58th session, A/61/10, forthcoming in YILC, commentary to Principle 2, para. 2.

[624] Lefeber 1996, p. 88.

[625] Birnie, Boyle & Redgwell 2009, p. 187; Lefeber 1996, pp. 86-87.

[626] IPCC *AR4, Working Group I Report: The Physical Science Basis*, Summary for Policymakers (2007), pp. 13-14; see also Chapter 1.

[627] 2006 Principles on the Allocation of Loss in the Case of Transboundary Harm Arising out of Hazardous Activities, ILC Report on the work of its 58th session, A/61/10, forthcoming in YILC, commentary to Principle 2, para. 3.

[628] *Cancun Agreements: Outcome of the Work of the* Ad Hoc *Working Group on Long-Term Cooperative Action under the Convention*, UNFCCC Decision 1/CP.16 (2010), para. 4.

general and in challenging a single major emissions source in terms of its contribution to global warming. The potential of the obligation to prevent significant transboundary harm can be inferred from Micronesia's challenge to the Prunéřov plant modernization plan. Micronesia accepted the fact that Prunéřov's share of the global GHG emissions was 'only' 0.0161 per cent and that the plant did not 'directly cause sea-level rise, change weather patterns and increase storms.'[629] It pointed out, however, that there were 5000 lignite-fired power plants worldwide and each of them accelerated climate change by contributing to total global CO_2 emissions. In fact, Micronesia demanded a transboundary EIA precisely because Prunéřov's serious environmental impacts could affect its territory, if only by contribution.

5.2.3.5 Causal Link

In bringing a claim based on an allegation of breach by the respondent state of the obligation to prevent significant transboundary harm through the emission of GHGs, the claimant state would have to overcome a number of legal obstacles, and causal proof may present a particularly large challenge. In being contingent on the occurrence of actual or potential harm, the obligation to prevent significant transboundary damage requires that there be a causal link between GHG emissions and climate change damage.[630] Assuming the existence of a general causal link between anthropogenic GHG emissions and global warming (see Chapters 1-2), the claimant state would have to show that (a) there is a causal connection between global warming and particular damaging effects of climate change and (b) to what extent GHG emissions originating from the respondent state are responsible for global warming, *i.e.* the extent to which global warming can be *attributed* to a given state. It may be difficult to attribute global warming to a particular state because the contribution of other states would have to be taken into account. Since at present not a single state is carbon-neutral,[631] the contribution of the claimant state would have to be considered too. Further, it is submitted that the exact

---

[629] Viewpoint of the Federated States of Micronesia on the complex renovation of Prunéřov II power plant 3x250 MWe plan, letter of 4 January 2010 from the Director of the Office of Environment and Emergency Management of the Federated States of Micronesia to the Ministry of the Environment of the Czech Republic.
[630] Lefeber 2012, p. 338; see also Lefeber 1996, p. 89.
[631] So far only the Maldives has announced its intention of becoming carbon-neutral by 2020; pledges of carbon neutrality have also been made by Bangladesh, Barbados, Bhutan, Ghana, Kenya, Kiribati, Nepal, Rwanda, Tanzania, and Vietnam, see Declaration of the Climate Vulnerable Forum, Malé, Maldives (10 November 2009), available from: <http://daraint.org/wp-content/uploads/2010/12/Declaration-of-the-CVF-FINAL2.pdf>.

standard of proof would vary depending on whether actual or potential harm is alleged as well as on the remedy sought.

For instance, if the respondent state is challenged for failure to prevent *actual* harm, the claimant state could seek reparation in the form of compensation as in the context of climate change restitution would likely be impossible (see Section 5.4). As described in Chapter 3, proof of a causal link between the wrongful act of a state and the damage suffered is necessary in order for the obligation to make reparation for injury to arise (see also Section 5.4).[632] With regard to climate change-related damage, the claimant state would have to prove that the respondent state's failure to take adequate mitigation measures has led, or contributed to, climate change. The claimant state would then need to prove the causal connection between GHG emissions originating in the respondent state and their contribution to climate change on the one hand, and the link between climate change and a particular event, which caused the harm complained of, on the other. Given the current state of the climate change science (see Chapter 1) and the global acceptance of the linkage between anthropogenic emissions and the warming effect (see above), proving that the emission of GHGs in a particular state contributes to causing the climate to change would not be difficult. However, showing that climate change has caused a particular extreme weather event, such as a hurricane or drought, to occur may present considerable problems. Proving causation in this instance would place on the claimant state a serious evidentiary burden as it would require the admission of statistical evidence, which has not been the practice of international courts.[633] The study of the practice of domestic courts in dealing with allegations of actual harm in claims for compensation has also revealed considerable causal proof difficulties (see Section 4.2.3.3). This finding could be indicative of a general principle of law and, as such, could be seen as corroborating the conclusion that proving the causal link for the purposes of obtaining compensation for actual harm may be an insurmountable hurdle also for a claimant state.

It is submitted that it may be easier for the claimant state to establish a causal connection between climate change and a slow onset event. The types of harm that result from slow onset events include, *inter alia*, loss of sovereign territory due to the rising sea levels, loss of fisheries to ocean acidification, and loss of productive land due to desertification. Slow onset events are gradual in occurrence, often permanent in character, and the link between such events and the warming of the

---

[632] The suitability of Satisfaction in the climate change context is also dealt with in Section 5.4.

[633] *E.g.* neither the ICJ Rules of the Court nor the PCA Rules of Procedure contain any provisions on the admissibility of statistical evidence.

climate system is supported by science.[634] However, due to the rigours of the *causa proxima* test required for the obligation of reparation to arise, it may still be a challenge (see Section 5.4). Lack of success in litigating compensation claims domestically seems to point in the same direction as plaintiffs have been unable to overcome causal challenges regardless of whether the harm complained of was due to an extreme weather event (*e.g. Comer*) or a slow onset event (*e.g. Kivalina*).

Additionally, even if the claimant state is able to show that GHG emissions from the respondent state contribute to global warming, which, *inter alia*, causes the sea levels to rise, which, in turn, leads to loss of territory in the claimant state, retroactivity may pose additional challenges as the law of state responsibility provides a guarantee against retroactive application of international law.[635] Due to the cumulative effects of climate change, the claimant state would have to demonstrate that all of the respondent state's emissions since the Industrial Revolution must be taken into account; consideration of GHG emissions released since the time of availability of scientific evidence on global warming alone would be insufficient. The IPCC AR4 has shown that the global effect of human activities dates back to the 1750s and that that effect has been one of warming.[636] The international community is in agreement on that as well, as is evidenced by the universal ratification of the UNFCCC.[637] To avoid the assumption of retroactive liability, the respondent state would likely propose introducing a cut-off date so that only the emissions from a particular time period are taken into account.[638] The respondent state could argue that it could only be held responsible for the harm caused after it knew or could have been expected to know about the connection between GHG emissions and the injurious effects of climate change. For instance, 1992, the year the UNFCCC was agreed upon, could serve as such a date[639] or 1994, the year the Convention entered into force, or the year the UNFCCC entered into force for the party concerned.[640] Recalling the gradual and cumulative character of climate change, it could be difficult to conclude that GHG emissions released between the cut-off date and the present time have alone been capable of causing global warming and the harm stemming from it. For about 200 years since the beginning of the Industrial Revolution, anthropogenic GHG emissions did not cause any major problems but recent emissions may have pushed the Earth's climate over the threshold. The claimant state could thus argue that the role of incremental emissions should be taken into account. Even so, given the prohibition

---

[634] *E.g.* see IPCC *AR4, Synthesis Report* (2007), p. 30.

[635] 2001 Articles on Responsibility of States for Internationally Wrongful Acts, Art. 13.

[636] IPCC *AR4, Synthesis Report* (2007), p. 37.

[637] See, in particular, reference to 'historical global emissions' in the 1992 UNFCCC preamble.

[638] See also Chapter 4 on cut-off dates in domestic claims.

[639] Farber 2008, p.10524.

[640] See Chapter 3.

against retroactivity and the difficulties with proving causation, it appears unlikely that an interstate claim for compensation for actual harm suffered can be sustained. Domestic practice, too, appears to be in line with this finding. Albeit courts have not explicitly dealt with retroactivity in the climate change context, they have indicated that they perceived the issue as problematic, and there is little doubt as to the general-principle-of-law status of the prohibition against retroactivity.[641]

Yet, historical counterarguments against retroactivity cannot be used to avoid responsibility altogether. The claimant state may choose to base its claim on breach of the obligation to prevent *potential* harm and seek cessation of wrongful conduct. It has been argued that in practice, cessation would amount to injunctive relief. In other words, to cease non-compliance with the obligation to prevent transboundary harm, the respondent state would need to start taking adequate mitigation measures, thereby exercising due diligence required of it by the obligation to prevent. In assessing the adequacy of mitigation measures, the principle of common but differentiated responsibilities and respective capabilities would need to be taken into account. For example, the level of adequacy of mitigation measures for a small developing state would be different from that to be expected of the US.

It would suffice for the claimant state to show that GHG emissions originating in the respondent state are *capable* of causing harm through a slow onset event or an extreme weather event. Thus, the evidentiary hurdles associated with proving causation between global warming and extreme weather events could also be overcome. The claimant state would have to prove that GHG emissions from the respondent state are, by way of their contribution to global warming, capable of: causing its territory to disappear because of the rising sea levels, causing its land to become unproductive due to desertification, or causing damage to its property through a hurricane or a flood. In other words, the claimant state would need to show that the respondent state's emissions are capable of causing foreseeable harm in the future. IPCC reports could be relied upon for arguing that the various types of harm caused by slow onset events or extreme weather events alike are objectively foreseeable. This finding is not only relevant to slow onset events but also to extreme weather events because it appears to suggest that the standard of proof for demonstrating the causal link in injunctive relief claims is lower than that in compensation claims. It should be noted that domestic claimants for injunctive relief have enjoyed a fair amount of success with proving causation. Although claims for injunctions have yet to be decided on the merits, successful demonstration of causation-as-an-element-of-standing seems to be indicative of a

---

[641] See *e.g.* 2001 Articles on Responsibility of States for Internationally Wrongful Acts, ILC Report on the work of its 53rd session, A/56/10, YILC, vol. II, Part Two, commentary to Art. 13, para. 4.

standard of proof that is lower than the one involved in compensation claims based on allegations of actual harm.

The claimant state could also allege a breach by the respondent state of certain procedural duties stemming from the obligation to prevent transboundary harm, *i.e.* the duty to prepare an EIA of a particular project, the duty to inform the states likely to be affected or the duty to consult and negotiate with the affected states. Micronesia's challenge has demonstrated that although a single GHG emissions source such as Prunéřov cannot be considered to cause sea level rise or increase storms directly, its contribution to the global emissions is sufficient to necessitate a transboundary EIAs, without which due diligence cannot be considered to have been exercised. This seems to suggest that in order to substantiate breaches of procedural duties, it is sufficient to show that the emissions from a particular source are capable, by way of contribution, of causing significant harm in the claimant state. This conclusion is also supported by numerous decisions from domestic jurisdictions. A careful consideration of procedural injury claims has shown that the causal link, both general and specific, has not been a major challenge for plaintiffs, and courts have frequently accepted it.

5.3 Origins of the International Responsibility of States for Climate Change-Related Damage

In the climate change context, states could incur international responsibility for breaches of their treaty obligations on mitigation as well as obligations under customary international law. The following section explores the possibilities of imposing international responsibility for climate change-related damage on heavily emitting states. Options presently available to the claimant state would differ depending on the origins of the international responsibility of the respondent state, *i.e.* the primary obligation breached, and on (a) whether or not the perpetrator state is a party to the Kyoto Protocol and, taking into account the principle of common but differentiated responsibilities, (b) whether the respondent state is an industrialized or a developing country. Section 5.3.1 addresses the origins of the international responsibility of industrialized states party to the Kyoto Protocol and Section 5.3.2 deals with the origins of the international responsibility of EITs. Section 5.3.3 focuses on the international responsibility of industrialized countries not party to the KP; and, finally, Section 5.3.4 describes the responsibility of developing states.

At the outset, the relationship between the customary obligation to prevent significant transboundary harm and the conventional obligations on mitigation must be clarified. In international law, whenever multiple norms deal with the same subject matter, the more specific norm is given priority. Assuming that the rationale behind the maxim *lex specialis derogat legi generali* is for the special rule to be more concrete in a particular context (climate change in the case at hand),[642] it is submitted that conventional international obligations on climate change mitigation do not constitute *lex specialis* in respect of the duty to prevent significant transboundary harm (see Section 3.2.2.1). Rather, the duty to prevent and international obligations on climate change mitigation coexist in parallel. The UNFCCC is a political framework agreement whose objective is to prevent dangerous anthropogenic interference with the climate system. The Convention does not impose any obligations on the participating states directed at the prevention of transboundary harm; it merely contains the common target of returning GHG emissions to 1990 levels by the end of the 20th century. The KP, it is argued, does not provide for more specific rules on harm prevention either. A political agreement, it spells out industrialized states parties' mitigation commitments in the form of quantified emissions limitation or reduction targets, which, again, are not directed at transboundary harm prevention. It is therefore submitted that states with quantified targets under the KP are also bound by the customary no-harm rule and breach of the duty to prevent can be measured against the state's compliance with the KP (see Section 5.3.1). This is also in line with the principle of harmonization, in accordance with which several norms bearing on the same issue should, insofar as it is possible, be interpreted as to give rise to a single set of compatible obligations, and general law will continue to apply in situations not provided for by the special rule.[643] Thus, the obligation to prevent significant transboundary harm will be binding on industrialized states with commitments under the KP.

5.3.1 Origins of the International Responsibility of Industrialized States Party to the KP

Under the international climate change regime, industrialized states have accepted more stringent commitments than those applicable to all states parties. Industrialized countries are bound by a wide range of substantive and procedural commitments

---

[642] *Conclusions of the work of the Study Group on the Fragmentation of International Law: Difficulties arising from the Diversification and Expansion of International Law*, ILC Report on the work of its 58th session, A/61/10, forthcoming in YILC (2006), para. 251(7).

[643] *Conclusions of the work of the Study Group on the Fragmentation of International Law: Difficulties arising from the Diversification and Expansion of International Law*, ILC Report on the work of its 58th session, A/61/10, forthcoming in YILC (2006), paras 251(4) and (9).

related to climate change mitigation and adaptation, which include enactment and coordination of policies and measures to mitigate climate change, achievement of quantified targets within a specific timeframe, consideration of specific needs of developing countries vulnerable to climate impacts in providing new and additional financial resources, and transfer of technology and capacity-building, and reporting all of the above in a transparent and verifiable manner.[644] The Kyoto Protocol prescribes individual quantified emissions limitation and reduction commitments for industrialized countries.[645] It also requires that, in achieving these commitments, developed countries implement and further elaborate policies and measures related to climate change mitigation.[646] Failure to live up to emissions limitation and reduction commitments will trigger the KP compliance procedure, which is designed not only to promote return to compliance, but also to ensure that the defaulting state does not take advantage of defaulting (see Section 2.2.2.2).

Canada provides an interesting example. When still party to the KP, it publicly declared that it did not intend to meet is emissions reduction target under the Protocol. Led by the Liberal government, Canada played an active role in the KP negotiations and committed itself to a 6 per cent target: it must reduce its emissions to 94 per cent of 1990 levels by 2013. The federal elections of 2006 changed the political climate in the country resulting in a minority government led by the Conservative Party. GHG emissions have been rising steadily and by 2007 increased 26.2 per cent above 1990 levels and 33.8 per cent above Canada's Kyoto target.[647] That year, the Canadian government unveiled 'Turning the Corner: An Action Plan to Reduce Greenhouse Gases and Air Pollution' that prescribed new mandatory industrial emissions reduction targets. The target set by the Action Plan was to reduce emissions by 20 per cent by 2020 and 60 to 70 per cent by 2050, using 2006 as the reference year.[648] Needless to say, this new goal was at odds with the Canadian reduction target under the KP. Although the Parliament rejected the

---

[644] 1992 UNFCCC, Art 4(2); see also F. Yamin & J. Depledge, *The International Climate Change Regime: a Guide to Rules, Institutions and Procedures*, Cambridge: Cambridge University Press, 2004, p. 105.

[645] 1997 KP, Art. 3.

[646] 1997 KP, Art 2.

[647] Canada's Fifth National Communication on Climate Change, submitted to the UNFCCC Secretariat on 12 February 2010, available from:
<http://unfccc.int/resource/docs/natc/can_nc5.pdf>, p. 21.

[648] Turning the Corner: An Action Plan to Reduce Greenhouse Gases and Air Pollution. Government of Canada news release of 26 April 2007, available from:
<www.ec.gc.ca/default.asp?lang=En&n=714D9AAE-1&news=4F2292E9-3EFF-48D3-A7E4-CEFA05D70C21> (last visited on 8 April 2011); see also Government of Canada news release of 12 December 2007, available from: <www.ec.gc.ca/default.asp?lang=En&n=714D9AAE-1&news=0F208D84-395E-4E78-8E6F-CCB906C30F5B> (last visited on 8 April 2011).

Action Plan and adopted the Kyoto Protocol Implementation Act, which is being implemented through a Climate Change Plan, Canada will still not achieve its KP reduction target.[649]

In 2008, the Canadian government's inaction was challenged in a domestic court by the local branch of the *Friends of the Earth* civil society organization.[650] The court, however, refused judicial review of the government's response to Canada's international obligations under the KP. More recently, the Canadian government announced its commitment to reduce its GHG emissions by 17 per cent below 2005 levels by 2020.[651]

On 15 December 2011, Canada withdrew from the Kyoto Protocol. The withdrawal took effect one year later pursuant to Article 27(2) of the KP, which provides:

> Any [...] withdrawal shall take effect upon expiry of one year from the date of receipt by the Depositary of the notification of withdrawal, or on such later date as may be specified in the notification of withdrawal.

Therefore, as of 15 December 2012, Canada was no longer bound by the KP provisions. The first commitment period under the KP ended on 31 December 2012, which means that Canada's withdrawal was effected only 16 days prior to that.

The timing of Canada's withdrawal raises a number of questions. First, it can be argued that Canada has acted in contravention of the *pacta sunt servanda* principle reflected in Article 26 of the Vienna Convention on the Law of Treaties, which provides that '[e]very treaty in force is binding upon the parties to it and must be performed by them in good faith.'[652] The fact that Canada withdrew from the KP two weeks before its obligation to meet its individual emissions target was due seems to point to the lack of *bona fide* intention to fulfil its emissions reduction obligations under the Protocol. In the alternative, Canada could have chosen (but did not) to purchase emission credits on the international carbon market to compensate for its growing emissions and thus achieve compliance. Second, Canada's late withdrawal from the KP raises proportionality concerns. Since Canada must be considered bound by the KP provisions until 15 December 2012, it

---

[649] See Climate Change Plan for the Purposes of the Kyoto Protocol Implementation Act - 2007, Environment Canada Inquiry Centre (2007), available from: <www.ec.gc.ca/doc/ed-es/p_123/CC_Plan_2007_e.pdf> (last visited on 8 April 2011).

[650] *Friends of the Earth – Les Ami(e)s de la Terre v. The Governor in Council and The Minister of the Environment,* 2008 FC 1183, 20 October 2008; see Chapter 4.

[651] Canada's Climate Change Mitigation Plan, presentation at workshop on clarification of the developed country parties' quantified economy-wide emission reduction targets and related assumptions and conditions, Bonn (17 May 2012), available from: <http://unfccc.int/files/bodies/awg-lca/application/pdf/20120517_canada_1749.pdf> (last visited on 23 December 2012).

[652] 1969 Vienna Convention on the Law of Treaties, Art. 26.

seems unbalanced that it should be altogether excused from its emissions reduction commitments – an obligation of result due by the end of a 4-year commitment period. It is submitted that, given its withdrawal, Canada must only be considered free from its Kyoto obligations for the last two weeks of 2012, and not for the entire 2008-2012 period. Therefore, Canada's original assigned amount target should be recalculated to reflect the two-week difference. This new target would only be fractionally lower, and Canada could still be held accountable under the law of state responsibility for its failure to meet it. Third, irrespective of whether or not Canada is considered to be bound by its revised quantified emissions reduction target that, as suggested earlier, would be due on 15 December 2012, it can be argued that it is still under an obligation to comply with the methodological and reporting requirements under the Protocol at least for the period preceding 15 December 2012 (see Chapter 2). This means that Canada would still be under an obligation to submit its annual national inventory containing information on GHG emissions, including methodologies used to estimate these emissions, as well as a national communication for 2012 (covering the period until 15 December 2012) containing, *inter alia*, information on: national GHG emissions and projections, mitigation policies and measures, vulnerability and adaptation to climate change, and financial assistance and technology transfer to non-Annex I parties.

Overall, the Compliance Committee EB can only assess the non-compliance of industrialized states with their targets under the KP after 2012 and, taking into account the KP reporting and review procedures, their non-compliance will not be established before 2015. The FB could exercise its early-warning function before the end of the first commitment period and provide advice to promote compliance.[653] In 2012, the FB tried to exercise its early-warning function with regard to Canada's potential non-compliance. On 9 February 2012, the FB chairperson sent a letter to Canada offering it an opportunity to respond to the concerns raised by the ERT. On 16 May 2012, Canada responded that on account of its notification of withdrawal from the KP, further engagement with the FB would be of little value, and the FB concluded its consideration of the matter.[654]

At present, it is impossible to determine whether an industrialized country has breached its emissions reduction obligations under the KP. Additionally, despite the laxity of domestic policies and measures and the government's expressed reservations, industrialized countries could still purchase emission credits on the international carbon market to ensure their ultimate compliance with the Protocol. The inventories for the last year of the first commitment period are not due before

---

[653] Lefeber & Oberthür 2010, p. 137; see also Chapter 2.

[654] *Annual report of the Compliance Committee to the Conference of the Parties serving as the meeting of the Parties to the Kyoto Protocol*, FCCC/KP/CMP/2012/6 (8 November 2012), paras 72-74.

15 April 2014, and ERTs will have up to one year to review them, which will be followed by a 100-day true-up period. During this additional period, developed countries party to the Protocol could still make use of the option to fulfil their commitments as they may 'continue to acquire, and other Parties may transfer to [them], emission reduction units, certified emission reductions, assigned amount units and removal units under Articles 6, 12 and 17 of the Protocol, from the preceding commitment period' provided their eligibility has not been suspended by the EB. [655] Therefore, an interstate claim to secure an industrialized state's compliance with the KP would have to be dismissed because the obligation to meet the reduction targets is not due before 2015.

Industrialized countries party to the KP, including Canada, could also be challenged for not living up to their customary duty to prevent significant transboundary harm. Canada is under an obligation to ensure that activities within its jurisdiction or control do not cause transboundary harm. It must exercise due diligence directed at significant transboundary harm prevention. It could be argued that upon becoming a party to the KP and undertaking quantified emissions reduction commitments, Canada implicitly recognized that reaching the GHG emissions reduction target would be equivalent to acting diligently in endeavouring to prevent significant transboundary harm resulting from climate change. Its current GHG emissions trajectory and recent withdrawal from the KP suggest that Canada is anything but exercising due diligence for the purposes of significant harm prevention. It is thus submitted that Canada has acted in breach its international customary duty to prevent significant transboundary harm.

Should Canada's endeavours to prevent climate change-related damage resulting from its GHG emissions be found inadequate under the due diligence standard, the next step would be to establish the causal link between Canada's emissions, climate change, its manifestations (*e.g.* the rising sea levels or extreme weather events), and the actual and/or potential damage because the duty to prevent is contingent on the actual or potential occurrence of harm. While, for the purposes of obtaining reparation, it would be relatively easy to prove that anthropogenic GHG emissions originating in Canada contribute to the global climate change, the causal link between climate change and some of its damaging effects, such as extreme weather events, is less straightforward (see Chapter 1). Establishing such a link may necessitate the admission of statistical evidence (see Section 5.2.3.5 above). The chances of substantiating a claim for reparation of Canada's breach of the obligation to prevent significant transboundary harm may be higher when the damage is manifested through the rising sea levels and ocean acidification as it

---

[655] *Procedures and mechanisms relating to compliance under the Kyoto Protocol*, UNFCCC Decision 27/CMP.1 (2005), Section XIII.

would be easier to prove that this type of damage is related to global warming (see
Chapter 1). Causal proof may be further complicated by the need to consider the
contribution of other states as well as the claimant state itself.

However, if cessation, or injunctive relief, is the appropriate remedy, it will
suffice to show that Canada's GHG emissions trajectory is merely capable of
causing significant transboundary harm, thus making the burden of proof less
onerous. In principle, all Canada's emissions from the time of the Industrial
Revolution would have to be taken into account, in which case retroactivity may
pose additional challenges. It can be argued, however, that the Earth's climate
system has been able to cope with the growing concentrations of GHGs in the
atmosphere for centuries whereas the increase in anthropogenic emissions of the
recent decades has tipped the balance (see Section 5.2.3.5).

In 2012, the KP first commitment period expired and, on 1 January 2013, the eight-
year second commitment period began. Same as the first commitment period, the
second commitment period only extends individual emissions limitation and
reduction commitments to industrialized states. Besides the US, at least two
developed states and one EIT – Canada, Japan, and the Russian Federation – have
undertaken no emissions limitation and reduction commitments under the KP
second commitment period.[656] New Zealand has indicated its intention to take on
an economy-wide emissions reduction target in accordance with the Convention,
and not the Protocol.[657] Yet, regardless of the level of participation of industrialized
states in the second commitment period, an interstate claim against them could still
be based on the customary rule to prevent significant transboundary harm.

At the moment it is unclear whether developing states will undertake binding
emissions reduction commitments under a new 'protocol, another legal instrument
or an agreed outcome with legal force under the Convention applicable to all
Parties' to come into effect in 2020[658] and whether, in the absence of such
commitments, all industrialized nations will come on-board. Irrespective of post-
2020 obligations on climate change mitigation, all states will remain bound by the
customary obligation to prevent significant transboundary harm. The duty to
prevent will continue to be binding even in the absence of specific treaty
obligations on climate change mitigation. Therefore, a potential interstate claim
could be based on customary international law whereas additional opportunities for

---

[656] *Outcome of the work of the* Ad Hoc *Working Group on Further Commitments for Annex I Parties under
the Kyoto Protocol at its sixteenth session*, UNFCCC Decision 1/CMP.7 (2011), Ann. 1.

[657] See *Outcome of the work of the* Ad Hoc *Working Group on Further Commitments for Annex I Parties
under the Kyoto Protocol*, FCCC/KP/CMP/2012/L.9 (UNFCCC decision number not available at the time of
writing), 8 December 2012, Ann. 1.

[658] *Establishment of an* Ad Hoc *Working Group on the Durban Platform for Enhanced Action*, UNFCCC
Decision 1/CP.17 (2011), para. 2.

holding states to account for climate change-related damage will depend on whether or not the climate negotiations produce a set of post-2020 legally binding mitigation commitments.

5.3.2 Origins of the International Responsibility of EITs

EITs, *i.e.* the Russian Federation and several other Central and Eastern European countries,[659] enjoy a special status under the Convention and the Protocol. However, since EITs are industrialized states and since they are also listed in KP Annex B (with the exception of Belarus, see below), the origins of their international responsibility would be largely similar to those of any other developed state party to the KP. Yet, their special position deserves separate mention.

Under the Convention, EITs are allowed 'a certain degree of flexibility' in implementing their commitments due to major economic and political upheavals that they went through following the demise of the Soviet Union.[660] Accordingly, several EITs have chosen a baseline earlier than 1990, *i.e.* before the economic changes which led to significant reductions in their GHG emissions.[661] Also, under the Convention, EITs, together with developing states, benefit from the obligation to transfer climate-friendly technologies imposed on Annex II parties.[662]

EITs are also entitled to 'a certain degree of flexibility' in implementing their commitments under the KP.[663] With the exception of Belarus, all EITs are included in Annex B of the KP.[664] The EB of the KP Compliance Committee interpreted the notion of 'a certain degree of flexibility' in the case concerning Croatia in 2009. With regard to emissions reduction or limitation commitments under the KP in the case concerning Croatia, the EB concluded that flexibility under Article 3(5) only extends to flexibility in the use of an historical base year or period other than 1990.[665] It further noted that 'a certain degree of flexibility' provided for under KP Article 3(6) refers to the implementation of commitments under the Kyoto Protocol

---

[659] Belarus, Bulgaria, Croatia, Czech Republic, Estonia, Hungary, Latvia, Lithuania, Poland, Romania, Slovakia, Slovenia, and the Ukraine.

[660] 1992 UNFCCC, Art. 4(6).

[661] Bulgaria, Hungary Poland, Romania, and Slovenia.

[662] 1992 UNFCCC, Art. 4(5).

[663] 1997 KP, Arts. 3(5), 3(6).

[664] As mentioned earlier, for Belarus, the amendment to Annex B concerning its emissions reduction target has not yet entered into force.

[665] Question of Implementation – Croatia, KP Compliance Committee, Final Decision, CC-2009-1-8/Croatia/EB (26 November 2009), Ann., para. 15(a).

other than those under Article 3.[666] Therefore, the relevant provisions under the KP provide no basis for allowing 'the addition of tonnes CO_2 eq to the level of emissions for a base year or period in the implementation of commitments under Article 3 of the Kyoto Protocol.'[667] In the case related to the Ukraine, the EB also noted that it could not grant flexibility and defer a decision or the application of consequences for EITs in the absence of a decision by the CMP that allows such flexibility under KP Article 3(6).[668]

It appears that, in practice, the flexibility in the implementation of commitments related to emissions reduction or limitation afforded by the KP to EITs only extends to the choice of a base year other than 1990; flexibility in the implementation of commitments other than those under KP Article 3 is subject to approval by the CMP. It should be added that under the Convention, EITs enjoy greater flexibility in meeting their commitments, and the degree of flexibility afforded to EITs by the CMP may also change in the future.

5.3.3 Origins of the International Responsibility of Industrialized States not Party to the KP

The US, the largest GHG emitter among industrialized nations, has not adopted a mitigation policy based on the KP.[669] At the time of the KP adoption, the US agreed to reduce its GHG emissions to 93 per cent of the 1990 levels but since it has not ratified the Protocol, it is not bound by this target. Its total aggregate emissions have grown by 13.3 per cent in the period between 1990 and 2008.[670] Although the US is not party to the Kyoto Protocol, it has ratified the UNFCCC and is bound by the obligation to adopt a national mitigation policy and take corresponding measures. The US government has adopted such policies and measures including a quantified target of reducing the emissions by 17 per cent below 2005 levels by

---

[666] Question of Implementation – Croatia, KP Compliance Committee, Final Decision, CC-2009-1-8/Croatia/EB (26 November 2009), Ann., para. 15(b).

[667] Question of Implementation – Croatia, KP Compliance Committee, Final Decision, CC-2009-1-8/Croatia/EB (26 November 2009), Ann., para. 15(c).

[668] Question of Implementation – Ukraine, KP Compliance Committee, Final Decision, CC-2011-2-9/Ukraine/EB (12 October 2011), para. 5(c).

[669] There are several other UNFCCC Annex I parties without KP Annex B targets. Due to its special circumstances recognized by the COP, Turkey is included in Annex I but not in Annex B. For Belarus, the amendment to Annex B concerning its emissions reduction target has not yet entered into force. Kazakhstan is party to the KP but does not have an emissions reduction target. Cyprus and Malta are EU member states but are not parties to the KP with a commitment inscribed in Annex B.

[670] *National Greenhouse Gas Inventory Data for the Period 1990-2008*, note by the secretariat, FCCC/SBI/2010/18 (4 November 2010), p. 11.

2020,[671] which could be seen as evidence of the US complying with its obligations on mitigation under the UNFCCC (see also Section 5.2.1). The US could be challenged, however, on the grounds that those measures are insufficient for it to comply with its customary obligation to prevent transboundary harm.

As a non-party to the Protocol, the US is not required to meet its KP quantified emissions reduction target. Instead, it may choose an alternative course of action to ensure compliance with its customary obligation to prevent transboundary damage. In June 2009, a comprehensive energy bill was passed by the House of Representatives by a vote of 219-212. American Clean Energy and Security Act 2009 includes provisions on clean energy, reducing global warming pollution, establishing a cap-and-trade system for GHG emissions, and transitioning to a clean energy economy. It also sets a reduction goal for GHG emissions from covered sources at 83 per cent of 2005 levels by 2050.[672] Additionally, the Act contains provisions promoting international reductions in industrial emissions and states that the US policy in this respect is 'to work proactively under the United Nations Framework Convention on Climate Change, and in other appropriate fora, to establish binding agreements, including sectoral agreements, committing all major greenhouse gas-emitting nations to contribute equitably to the reduction of global greenhouse gas emissions.'[673] The Act imposes an international reserve allowance requirement on imported goods per industrial sector in order to prevent carbon leakage in the event the US does not enter into a binding climate change agreement. Under the international reserve allowance program, countries that have taken adequate measures to mitigate the carbon intensity of a particular sector may receive emission allowance rebates whereas states that have not taken such measures are subject to the international reserve allowance requirement.[674] The Act, however, remains a legislative proposal as the Senate has not yet taken a vote on it, nor is it expected to any time soon.

Like states that have ratified, but not complied with, the KP and acted in breach of due diligence,[675] the US could be challenged for acting in breach of its international

---

[671] US Climate Action Report 2010, Fifth National Communication on Climate Change of the United States of America under the UNFCCC, p. 3, available from:
<http://unfccc.int/resource/docs/natc/usa_nc5.pdf >.
[672] American Clean Energy and Security Act 2009, H.R. 2454 (111th Congress), available from: <http://democrats.energycommerce.house.gov/Press_111/20090701/hr2454_house.pdf> (last visited on 15 June 2011).
[673] American Clean Energy and Security Act 2009, H.R. 2454 (111th Congress), Sec. 765, p. 1113.
[674] See also Section 3.2.2.4.1 on countermeasures and retortions and Section 5.5.3 on countermeasures.
[675] Not complying with KP targets may automatically mean that due diligence has also been breached. Whether a state has acted with due diligence must be determined on a case-by-case basis.

customary obligation to prevent significant transboundary harm. Bound by international custom, the US is under an obligation to ensure that activities within its jurisdiction or control do not cause significant transboundary harm. It must act diligently in its endeavours to prevent such harm, which in the climate change context implies that it must take adequate mitigation measures. As detailed in Section 5.2.3.5 above, it is sufficient for the claimant state to demonstrate that the US's failure to take adequate mitigation measures is only capable of causing climate change-related damage, such as damage from slow onset events or extreme weather events. Yet, since the UNFCCC does not specify what mitigation measures are to be considered adequate, it may be difficult to prove that the measures that are currently being taken by the US are insufficient to meet the due diligence threshold.

It could be argued, however, that deferral of federal action on climate change and GHG emissions reduction, coupled with the emissions trajectory that is inadequate to help keep the global temperature increase below 2°C, amounts to a breach of the US's customary obligation to prevent significant transboundary damage. Stalling clean energy legislation, it could be claimed, shows reluctance of the US government to adopt measures that conform to a higher standard of diligence than the country's current climate policy. It could be argued that since the due diligence threshold shifts in accordance with scientific progress, mitigation measures reported on by the American government in 2006 should have been upgraded pursuant to the new developments in the field of climate change science published in the IPCC FAR in 2007. A claim alleging the US's breach of its obligation to prevent significant transboundary harm could also face causation and retroactivity challenges similar to the ones described in the Canadian example above.

Regardless of whether or not a new mitigation agreement is concluded (see Chapter 2), the international responsibility of the US could still originate from breaches of the customary obligation to prevent significant transboundary harm (see Section 5.3.1 as the origins of state responsibility under custom are the same regardless of KP ratification).

5.3.4   Origins of the International Responsibility of Developing States

It has been pointed out that the Kyoto Protocol imposes no quantified emissions reduction or limitation targets on developing states.[676] Following the principle of common but differentiated responsibilities and respective capabilities, the international climate regime places developing states under no obligation to adopt mitigation policies and measures. It is accepted that the developing countries' share

---

[676] See Section 5.2.1.

of global emissions will grow and their right to development is accordingly recognized. [677] Developing states are only required to formulate national programmes containing measures to mitigate climate change but the implementation of this obligation by developing states is contingent on the provision of financial and technological resources by industrialized countries. [678]

Currently, the developing states' contribution to the global emissions is swiftly rising because of rapid population and economic growth those countries are experiencing. As stipulated in the Cancun Agreements, developing countries have been invited to take nationally appropriate mitigation actions, and many of them have done so, but in and of themselves those commitments have no legally binding effect (see Section 5.2.1). Some developing countries have made use of this option. For instance, China – the world's largest emitter of GHGs – has informed the UNFCCC Secretariat of its intention to lower its carbon dioxide emissions per unit of GDP by 40-45 per cent by 2020 compared to the 2005 level. [679] However, since developing countries are not legally bound to take NAMAs, they cannot be held responsible for their failure to meet those voluntary targets.

Notwithstanding the principle of common but differentiated responsibilities and respective capabilities, developing countries are required to prevent significant transboundary harm because this customary obligation is equally binding on industrialized and developing nations. Developing states must conduct themselves with due diligence in their efforts to prevent such harm. A claimant state could argue that in discharge of their due diligence duty of harm prevention, developing countries must take some sort of mitigation action despite the fact that they are not specifically required to do so under the UNFCCC. It could be reasoned that formulating a national programme containing measures to mitigate climate change cannot be considered sufficiently diligent unless some of those measures are actually implemented.

It could be argued that all states should be subject to a minimum level of due diligence, with any additional diligence being voluntary. Yet, it could also be argued that developing states cannot be expected to exercise the same degree of due diligence as industrialized nations. Also, the level of diligence required of a particular developing country cannot be lower than that required of other countries with an equivalent level of prosperity. [680] Seeing that the level of prosperity of

---

[677] See 1992 UNFCCC, preamble.

[678] 1992 UNFCCC, Art. 4(7).

[679] Letter including autonomous mitigation actions from the Department of Climate Change, National Development and Reform Commission of China, to the UNFCCC Secretariat of 28 January 2010, available from: <http://unfccc.int/files/meetings/cop_15/copenhagen_accord/application/pdf/chinacphaccord_app2.pdf>.

[680] See, *e.g.* ILC Report on the work of its 50th session, A/53/10, YILC, vol. II, Part Two (1998), commentary to Art. 3, p. 28, para. 14; see also Section 5.2.3.1.

certain developing countries is significantly higher than that of other developing nations (and even some industrialized states), the level of due diligence expected of developing states with a higher level of prosperity cannot be the same as the level of due diligence expected of developing states with a lower level of prosperity. What is of relevance here is not the developed/developing dichotomy inherent in the international climate regime but rather the individual circumstances of a particular state, the level of scientific knowledge and technical capability available to it as well as its historic contributions.[681] Due diligence must be assessed on a case-by-case basis. Thus, a developing state's efforts, *e.g.* those of China, in exercising due diligence towards significant transboundary harm prevention as reflected in its NAMA's voluntary mitigation target could, in principle, be challenged as insufficient. First, it could be argued that China's target does not reflect its level of scientific and technological development. Second, it could be claimed that its NAMA is not ambitious enough to be considered as an adequate contribution to the international community's effort to keep the global temperature rise below 2°C.

The claimant state could also challenge China by invoking the obligation to prevent significant transboundary harm in contesting China's failure to comply with certain substantive and/or procedural obligations inherent to the due diligence element of the preventive duty. For example, notwithstanding the principle of common but differentiated responsibilities and respective capabilities, the claimant state potentially affected by climate change could demand that China comply with those duties in a similar way as Micronesia challenged the modernization plan of a Czech power plant. The claimant state could demand that China make an EIA taking into account transboundary impacts of a particular project, issue a notification to potentially affected states and/or continue monitoring transboundary environmental impacts of a specific activity throughout its operational cycle. The claimant state could also argue that, at the national level, China must take 'reasonably appropriate' mitigation action consistent with its level of technological development, adopt appropriate rules and measures, and exercise 'a certain level of vigilance in their enforcement and the exercise of administrative control applicable to public and private operators.' [682] In the absence of such measures and their enforcement, due diligence associated with significant transboundary harm prevention would not be considered to have been fulfilled.

---

[681] See *Responsibilities and Obligations of States with Respect to Activities in the Area*, Advisory Opinion of 1 February 2011, 2011 ITLOS Rep. 10, para. 162.

[682] See *Case Concerning Pulp Mills on the River Uruguay (Argentina v. Uruguay)*, Judgment of 20 April 2010, 2010 ICJ Rep. 14, at 79, para. 197; *Responsibilities and Obligations of States with Respect to Activities in the Area*, Advisory Opinion of 1 February 2011, 2011 ITLOS Rep. 10, para. 120.

Should developing states undertake binding commitments under a new mitigation agreement, their international responsibility could originate from possible breaches of treaty obligations. However, having repeatedly emphasized the NAMAs' voluntary character, developing countries may not wish to commit to any legally binding emissions reduction or limitation targets. Therefore, breach of the customary obligation to prevent significant transboundary harm may provide the sole legal basis for invoking the international responsibility of states, industrialized and developing alike, for taking inadequate mitigation measures.

5.4 Content of State Responsibility for Climate Change-Related Damage

The present chapter has identified the primary obligations of states under the international climate regime as well as international custom distinguishing between industrialized countries and EITs party to the KP, industrialized countries and EITs not party to the KP, and developing countries. Further, the origins of state responsibility for each group of states have been examined. In situations when a primary obligation is breached and the international legal order is impinged upon by the commission of an internationally wrongful act, states enter into a special legal relationship governed by the secondary rules of state responsibility. According to those rules, certain states derive rights from such an impingement whereas the perpetrator state must repair the breach. If an internationally wrongful act originates from a serious breach of a peremptory norm of international law, other states are duty-bound to respond to a breach of such an obligation (see Section 3.2.2.3.2).

The perpetrator state is under an obligation to cease the wrongful act if it is of a continuous nature and, in some situations, offer guarantees of non-repetition. Thus, it is submitted that the law of state responsibility could provide a legal basis for seeking injunctive relief that would require the responsible state to take adequate mitigation measures or discontinue the construction or modernization of a single major emitting source until the relevant procedural and substantive duties are fulfilled.[683] Two essential conditions for the obligation of cessation to arise have been stressed in the Rainbow Warrior arbitration, 'namely that the wrongful act has a continuing character and that the violated rule is still in force at the time in which the order is issued.'[684] It is apparent that interference with the climate system is a lengthy process that can only be achieved through an act of a continuous nature.

---

[683] See Chapter 3; for national claims seeking injunctive relief, see Section 4.2.2.

[684] 2001 Articles on Responsibility of States for Internationally Wrongful Acts, ILC Report on the work of its 53rd session, A/56/10, YILC, vol. II, Part Two, commentary to Art. 30, p. 217, para. 3, in reference to Rainbow Warrior (New Zealand v. France), 30 April 1990, UNRIAA, vol. XX, p. 215 (2006), p. 270, para. 13.

Failure to take adequate mitigation measures at the state level could qualify as an act of continuous nature interfering with the climate system.

It could be argued that by not taking adequate mitigation measures, the respondent state has not been diligent in endeavouring to prevent significant transboundary harm. Thus, failure to take adequate mitigation measures could amount to a violation of the obligation to prevent. Cessation of wrongful conduct in this context implies that the respondent state would have to start taking such measures. Thus, for the claimant state to demand cessation of wrongful conduct would in effect be equivalent to seeking an injunction against the perpetrator state. The claimant state would have to demonstrate that failure by the respondent state to take adequate mitigation measures has the *potential* of causing significant transboundary harm (see Section 5.2.3.5). Since no *actual* harm is necessary for demonstrating breach of the primary norm, the standard of causal proof is lower than that required for a successful claim for reparation (see below).

Also, the obligation of cessation offers the basis for challenging a single major emitting source in the respondent state and demanding compliance with procedural and/or substantive obligations inherent in the duty to prevent transboundary harm. In this context, cessation of wrongful conduct would amount to compliance with the relevant procedural and substantive duties, *e.g.* making an EIA where one has not been made or, like in Micronesia's challenge of a power plant modernization project in the Czech Republic, demanding that a project's EIA take into account its transboundary impacts. In terms of causation, it would suffice for the claimant state to show that a particular GHG emitting source is capable of causing it *potential* harm, which is foreseeable (see also Section 5.2.3.5).

Cessation and non-repetition are directed at the re-establishment of the legal relation disrupted by an internationally wrongful act but on their own they do not fully restore the international legal order. Therefore, the responsible state is required to eliminate the consequences of the wrongful act by making reparation, which must 'as far as possible, wipe out all the consequences of the illegal act and re-establish the situation which would, in all probability, have existed if the act had not been committed.'[685] Reparation can take the form of restitution, which can be achieved through clean-up or response measures; compensation or satisfaction. Reparation is subject to the requirement of a causal connection between the internationally wrongful act and the damage suffered. An analysis of the reparation modules available under the law of state responsibility suggests that, in the climate change context, restitution in kind would likely be impossible. Satisfaction may play an important role as declaratory judgments have been made 'to ensure respect

---

[685] *Factory at Chorzów (Germany v. Poland)*, Merits, Judgment 1928 PCIJ (Ser. A) No. 13, p. 47; see also Chapter 3.

of international law.'[686] Compensation, however, may be a more attractive option for the claimant state.

The legal consequences described above arise automatically once the perpetrator state has committed an internationally wrongful act; other states are not required to take any action to trigger those consequences. Yet, the claimant state is only entitled to reparation if there is a causal link between the injury and the internationally wrongful act.[687] In international law, there are two causality tests: (1) the *conditio sine qua non* test and (2) the *causa proxima* test.[688] According to the *conditio sine qua non* test, the harm would not have occurred but for the event causing it. This test requires that there be a factual relation between the cause and the harm sustained. The number of links in the causal chain is irrelevant as long as the chain is uninterrupted. Causality-in-fact is a necessary but not a sufficient condition for reparation.[689] The more stringent *causa proxima* test requires that the harm be not too remote from the cause; the damage must be its normal or natural consequence. The ILC associates the proximate cause test with the exclusion of injury that is too 'remote' or 'consequential' and refers to the criteria of 'directness,' 'foreseeability,' and 'proximity.'[690] The foreseeability criterion suggests that the respondent state foresaw or should have foreseen the harm caused by its conduct. In other words, the harm must be objectively foreseeable. For instance, since compliance with the obligation to prevent transboundary harm is determined by a due diligence standard, foreseeability becomes a requirement for proving causation.[691] The claimant state would have to prove that the respondent state foresaw, or should have foreseen, that its unmitigated or inadequately mitigated GHG emissions, by way of their contribution to climate change, would have caused damage to the claimant state through the injurious manifestations of climate change, such as an extreme weather event or the rising sea levels. Given the universal adherence to the UNFCCC, it would not be difficult to prove that the respondent state foresaw the contributory effects of its GHG emissions on the global climate. As far as the link between global warming and the damage caused through its manifestations is concerned, it has been observed earlier in the chapter that linking

---

[686] *Corfu Channel (United Kingdom of Great Britain and Northern Ireland v. Albania)*, Judgment of 9 April 1949, ICJ Rep.4, at 35.

[687] 2001 Articles on Responsibility of States for Internationally Wrongful Acts, ILC Report on the work of its 53rd session, A/56/10, YILC, vol. II, Part Two, commentary to Art. 34, p. 235, para. 1.

[688] Lefeber 1996, pp. 89-93.

[689] 2001 Articles on Responsibility of States for Internationally Wrongful Acts, ILC Report on the work of its 53rd session, A/56/10, YILC, vol. II, Part Two, commentary to Art. 31, p. 227, para. 10.

[690] 2001 Articles on Responsibility of States for Internationally Wrongful Acts, ILC Report on the work of its 53rd session, A/56/10, YILC, vol. II, Part Two, commentary to Art. 31, pp. 227-228, para. 10.

[691] Lefeber 1996, p. 96.

the emission of GHGs originating from a particular state to damage resulting from the rising sea levels or other slow onset events in another may be less problematic than connecting it to the harm caused by an extreme weather event (see Section 5.2.3.5). Successful invocation of state responsibility in the latter case may be impeded by the need to introduce statistical evidence. In any event, since the *causa proxima* test required for the obligation of reparation to arise is stricter than the one used to determine breach of the obligation to prevent significant transboundary harm, making a successful compensation claim may not be easy. Since the *causa proxima* test requires that the damage be the normal or natural consequence of the cause, the state claiming compensation would not only have to show the causal link between GHG emissions originating in the respondent state and global warming. It would also have to prove that a particular event that has caused it injury was the natural consequence of global warming. This, particularly with respect to extreme weather events, may present considerable difficulties. Retroactivity, too, may pose additional challenges in compensation claims (see Section 5.2.3.5).

Additionally, even if the court were to accept the claimant state's causation and retroactivity arguments, determining the due amount would not be without difficulty given the perpetrator state's responsibility for a mere proportion of the total damage. Given the notion that each state is separately responsible for conduct attributable to it and that its responsibility is not diminished by the fact that other states may be responsible for the same act, it could be argued that responsibility should be proportionate rather than joint and several (see also Section 5.5.1).[692] The ILC neither accepts nor rejects joint and several responsibility of states. Yet, in the context of climate change, joint and several responsibility would make little sense because all states, including the claimant state, are responsible for climate change and its injurious effects. It would seem reasonable for the respondent state to be held responsible in proportion to the amount of GHGs it has contributed.

Additional legal consequences apply to serious breaches of peremptory norms of international law. They have not yet fully crystallized into custom but the duty of states to cooperate in bringing such breaches to an end is certain.[693] Already in 1976, the ILC listed 'modifications of weather and climate' among serious breaches of international obligations 'of essential importance for the safeguarding and preservation of the human environment, such as those prohibiting massive pollution

---

[692] 2001 Articles on Responsibility of States for Internationally Wrongful Acts, ILC Report on the work of its 53rd session, A/56/10, YILC, vol. II, Part Two, commentary to Art. 47, pp. 314, 317, paras 1, 6.
[693] See Section 3.2.2.3.2.

of the atmosphere or of the seas.'[694] As is evidenced by the UNFCCC's universal ratification, the international community of states has accepted the scientific evidence of the fact that anthropogenic GHG emissions contribute to climate change. The conclusion must be that obligations on climate change mitigation are peremptory norms of international law. [695] No derogation from those obligations may be permitted. The capacity of climate change to destroy lives and livelihoods on a global scale is generally recognized by the international community; in fact, the injurious consequences of climate change threaten the very existence of entire countries and, in the worst-case scenario, the human race as we know it (see Chapter 1). Although the qualification of international obligations on climate change mitigation as peremptory norms of international law has not been extensively debated in academic literature, it cannot but logically follow that those obligations are peremptory norms.

A breach of a peremptory norm is serious if it involves 'a gross or systematic failure by the responsible state to fulfil the obligation.'[696] The ILC has defined 'systematic failure' as a violation 'carried out in an organized and deliberate way,' while the term 'gross' refers to the intensity of the violation and 'denotes violations of a flagrant nature, amounting to a direct and outright assault on the values protected by the rule.'[697] On the one hand, since climate change mitigation calls for long-term planning and implementation efforts, proof of systemic breach may be relatively straightforward. On the other hand, all states agree on the fact that anthropogenic climate change poses a threat to human survival and that it is necessary to mitigate its dangerous consequences; it is the exact content of obligations involving mitigation measures that is disputed, which, in the absence of an internationally agreed benchmark, may compromise proof of breach of obligations related to climate change mitigation in the first place unless absolutely no mitigation measures are taken. It is submitted that consideration of GHG emissions trajectories compared to the 2°C warming target may aid proof of breach of such obligations. In this respect, GHG inventories submitted by developed countries to the UNFCCC Secretariat and national communications submitted by developing countries could provide some guidance.

---

[694] 1976 Draft Articles on Responsibility of States for Internationally Wrongful Acts, ILC Report on the work of its 28th session, A/31/10, YILC, vol. II, Part Two, Art. 19(3)(d), p. 96 and commentary to Art. 19, p. 109, para. 32.

[695] Lefeber 2012, pp. 342-343.

[696] 2001 ILC Articles on Responsibility of States for Internationally Wrongful Acts, Art. 40(2).

[697] 2001 Articles on Responsibility of States for Internationally Wrongful Acts, ILC Report on the work of its 53rd session, A/56/10, YILC, vol. II, Part Two, commentary to Art. 40, p. 285, para. 8.

5.5 Implementation of State Responsibility for Climate Change-Related Damage

Notwithstanding the fact that the consequences of an internationally wrongful act arise automatically upon its commission, other states have the right to hold the perpetrator state accountable for its wrongful conduct.[698] They may launch interstate claims and trigger procedures to enforce those claims through diplomatic negotiations, third-party dispute settlement and/or unilateral measures. The rules on the implementation of state responsibility spell out conditions this right is subject to. The notion of an injured state is central to the law of state responsibility; it is only in exceptional cases that a state other than the injured state can invoke the responsibility of a perpetrator state.

5.5.1   Invocation of State Responsibility

As discussed in Chapter 1, certain effects of climate change may be beneficial for some states, at least in the short term. Conversely, other states, such as small island nations and states with a low coastline, are already enduring the injurious effects of climate change. Coastal territories and certain islands are becoming uninhabitable, due to storm surges intensity and increasing damage from coastal flooding,[699] and will have to be permanently abandoned even before they are submerged completely. This may eventually lead to a sovereign state having to abandon its territory and relocate its entire population. One way or another, all states on the planet will be affected by climate change. Already now a number of states would have no difficulty demonstrating climate change-related injury and, consequently, *locus standi* to bring an interstate claim. The plurality of injured states does not affect the right of each individual state to invoke responsibility.[700] The plurality of responsible states is likewise not an obstacle as the responsibility of each contributing state may be invoked.[701] Domestic practice also appears to suggest that alleging injury should not be an arduous task. At the national level, standing, albeit not an obstacle in procedural injury cases, has been difficult to show for plaintiffs bringing claims for compensation or injunctive relief, alleging actual or potential harm, respectively. However, standing

---

[698] While the present analysis is limited to states as subjects of international law, international organizations may also have the right to invoke state responsibility. For more information, see 2011 ILC Draft Articles on the Responsibility of International Organizations, ILC Report on the work of its 63rd session, A/66/10, forthcoming in YILC, especially Art. 43.

[699] See Chapter 1; see also IPCC *AR4, Synthesis Report*, pp. 33, 52.

[700] See Section 3.2.2.4; see also 2001 ILC Articles on Responsibility of States for Internationally Wrongful Acts, Art. 46.

[701] See Section 3.2.2.4; see also 2001 ILC Articles on Responsibility of States for Internationally Wrongful Acts, Art. 47(1).

has not been challenging for lack of injury. Whereas, in showing standing, plaintiffs in US courts have struggled to meet the requirements of causation and redressability, demonstrating injury has not been problematic.

Since, as explained earlier, no state is carbon-neutral, all countries contribute to climate change, and the international community of states has agreed to tolerate this. An injured state could only invoke the responsibility of another state (or multiple states) if that state's contribution to climate change has exceeded the tolerance threshold marked by the 2°C goal. It could be argued that by not taking adequate mitigation measures, the perpetrator state has failed to keep its contribution to climate change within those tolerated boundaries, thus failing to prevent harm regarded as significant. A likely claimant would be a small island state suffering from the injurious effects of climate change. In fact, a number of such small island states, upon signing the UNFCCC, made a formal declaration that their respective governments' signature of the Convention 'shall in no way constitute a renunciation of any rights under international law concerning state responsibility for the adverse effects of climate change, and that no provisions in the Convention can be interpreted as derogating from the principles of general international law.'[702] While legally redundant, those declarations reaffirm existing international law by articulating the right of injured states to bring an interstate claim under the law of state responsibility. Nonetheless, no state suffering climate change-related injuries has made use of this right to date with the exception of Micronesia, who submitted its complaint through diplomatic channels. In 2002, the island nation of Tuvalu considered suing the US and Australia for their contribution to global warming but later abandoned the idea having weighed the difficulties associated with winning such a case. The government of Australia has since become a party to the KP while the US has not done so.

As regards the plurality of responsible states, the ILC neither recognizes nor excludes joint and several responsibility. The law of state responsibility is predicated on the principle of individual responsibility of states for their wrongful behaviour; the injured state can hold each responsible state to account 'for the wrongful conduct as a whole.'[703] Yet, in the absence of strict ILC standards for shared responsibility and due to the fact that climate change is gradually compounded by the emission of GHGs from multiple sources in multiple jurisdictions and the degradation of sinks worldwide over time, the application of proportionate responsibility to reparation claims would seem appropriate.[704] The GHG inventory data compiled by UNFCCC Secretariat could be relied upon in the proportionate assessment of states' contributions to climate change.

---

[702] Nauru, Tuvalu, the Republic of Kiribati, Fiji, and Papua New Guinea.

[703] 2001 Articles on Responsibility of States for Internationally Wrongful Acts, ILC Report on the work of its 53rd session, A/56/10, YILC, vol. II, Part Two, commentary to Art. 47, p. 314, para. 3 and p. 317, para. 6.

[704] Lefeber 2012, p. 346.

A state other than the injured state could invoke the responsibility of the respondent state only in exceptional circumstances, *i.e.* if the latter is in breach of an obligation *erga omnes*. It could be argued that although all states have an individual interest in other states' compliance with their obligations on climate change mitigation, those obligations also protect a collective interest and are therefore obligations *erga omnes* since climate change affects not only states but also territories outside the limits of national jurisdiction.[705] Obligations on climate change mitigation are also aimed at protecting the atmosphere – a shared natural resource.[706] A state other than the injured state could, on behalf of the injured state, demand that the responsible state take adequate mitigation action or, for example, it could challenge the modernization of a single emissions source. While not entitled to reparation itself, a state other than the injured state could demand that the obligations of cessation and reparation be performed in the interests of the beneficiaries of the obligation breached.[707] The ILC has recognized that although this provision involves 'a measure of progressive development,' it is justified because it 'provides a means of protecting the community or collective interest at stake.'[708] If climate change-related damage has been caused to areas beyond national jurisdiction, compensation could, in theory, be paid to an existing financial mechanism that sponsors mitigation or adaptation projects.[709] Admittedly, in the absence of precedent of such a claim being awarded by an international court or tribunal, international courts and tribunals would be unlikely to award such claims in the near future.[710]

5.5.2   Competent Courts

Closely related to the issue of invocation of state responsibility is the question of appropriate fora for third-party dispute settlement. Before bringing a claim to an international court, the invoking state would submit a diplomatic claim, following

---

[705] See 2001 ILC Articles on Responsibility of States for Internationally Wrongful Acts, Art. 48.

[706] 'Protection of the atmosphere' has now been included in the ILC Long-Term Programme of Work. See ILC Report on the work of its 63rd session, A/66/10, forthcoming in YILC (2011), p. 7, para. 32; on equitable and reasonable utilization of shared resources, see 1997 Convention on the Law of the Non-navigational Uses of International Watercourses, Doc. A/51/869 (11 April 1997), Art. 5; see also *Case Concerning the Gabčikovo-Nagymaos Project (Hungary v. Slovakia),* Judgment of 25 September 1997, 1997 ICJ Rep. 7, at 56, para. 85.

[707] 2001 ILC Articles on Responsibility of States for Internationally Wrongful Acts, Art. 48(2).

[708] 2001 Articles on Responsibility of States for Internationally Wrongful Acts, ILC Report on the work of its 53rd session, A/56/10, YILC, vol. II, Part Two, commentary to Art. 48, p. 323, para. 12.

[709] Cf: *Responsibilities and Obligations of States with Respect to Activities in the Area,* Advisory Opinion of 1 February 2011, 2011 ITLOS Rep. 10, paras 179-180 where the Tribunal recognized the right of all states party to the 1982 UNCLOS to claim compensation for breaches of obligations *erga omnes*; for claims for compensation in domestic courts see Chapter 4; see also Lefeber 2012, p. 347.

[710] Lefeber 2012, p. 347.

which negotiations would take place. Only after available bilateral procedures have been exhausted may states resort to third-party dispute settlement. The UNFCCC provides that parties to a dispute 'shall seek a settlement [...] through negotiation or any other peaceful means of their own choice.'[711] This means that although priority is clearly given to negotiation, other diplomatic means, such as good offices or mediation by a third party, complement negotiations.[712]

Should states fail to achieve a settlement, the state invoking the responsibility of another state may launch a claim in a permanent court or an *ad hoc* arbitral tribunal may be established for the settlement of a particular dispute. However, because of its binding character, international adjudication/arbitration is subject to acceptance of the competent court's jurisdiction by all parties to a dispute. The UNFCCC provisions on dispute settlement apply *mutatis mutandis* to the Kyoto Protocol and, next to diplomatic methods of settling a dispute, envisage mandatory recourse to non-binding conciliation and optional recourse to the ICJ and/or arbitration.[713] Compulsory adjudication is subject to optional declarations to be submitted by parties to the depositary. By submitting a declaration in accordance with the compromissory clause under the UNFCCC Article 14(2) and/or KP Article 19, parties may opt for compulsory adjudication and indicate their preference for submission of disputes to the ICJ, arbitration or accept both options. The effectiveness of this clause is uncertain because its applicability does not only depend on whether or not all parties to a dispute have accepted compulsory adjudication; it is also contingent on all parties' recognition of the jurisdiction of the same court. In practice, states parties have been reluctant to make use of the right to make optional declarations. Only three countries have done so to date. Solomon Islands and Tuvalu submitted upon ratification declarations choosing for compulsory arbitration and, in 2010, the Kingdom of the Netherlands made a declaration accepting both means of dispute settlement.

Acceptance of the UNFCCC/KP compromissory clause is not the only way for parties to a dispute to submit themselves to the jurisdiction of an international court. For example, they may recognize the ICJ's compulsory jurisdiction under Article 36 of its Statute or under any other instrument relating to amicable settlement of disputes. Only about one third of all states have accepted the ICJ's jurisdiction under the Statute of the Court. Major emitters like the USA and China have not done so. Canada, as has been noted earlier, has withdrawn from the KP and is on an emissions trajectory that may result in a breach of its international obligations, including customary duties. It did submit a declaration under the ICJ Statute in 1994, so in principle the Court could

---

[711] 1992 UNFCCC, Art. 14(1).
[712] Treves 2009, p. 501.
[713] 1992 UNFCCC, Art. 14; 1997 KP, Art. 19.

exercise jurisdiction over a claim invoking Canada's international responsibility provided that the invoking state has likewise recognized its jurisdiction.

Similarly, the claimant state and the respondent state could choose to submit their dispute to arbitration by a tribunal set up especially for that purpose or to a permanent arbitral body such as the Permanent Court of Arbitration (PCA) in The Hague. Recognition of the arbitral tribunal's jurisdiction by all parties to the dispute is a prerequisite due to the binding nature of its decisions.

The UNFCCC also provides for compulsory conciliation in cases when negotiation, or any other means the parties to a dispute decide to choose, does not lead to settlement within twelve months following notification by one party to another.[714] The award rendered by the conciliation commission created upon the request of one of the parties to the dispute is of a recommendatory nature (Art. 14(6)). The parties to a dispute are always entitled not to accept the solution suggested by the conciliation commission. Indeed, the main difference between arbitration and conciliation as envisioned by the UNFCCC is the non-binding character of the latter. However, effectiveness of the UNFCCC conciliation procedure must not be underestimated as it 'may serve as a powerful incentive for inducing the other Party or Parties to negotiate in good faith or to seek, in good faith also, to agree on other methods of settling the dispute.'[715]

5.5.3   Countermeasures

The conclusion from the previous section must be that the UNFCCC/KP dispute settlement rules, like under many other MEAs, are not all applicable to all the parties as, somewhat paradoxically perhaps, compulsory adjudication envisaged by those rules is in fact optional, *i.e.* is dependent on optional declarations by parties. Since collective enforcement mechanisms in international law are rare,[716] the 'practice whereby each state party to a treaty monitors whether or not other states parties comply with the requirements of the agreement' remains relevant.[717] In this context, self-help may be the only remaining option for the injured state to see to it that the legal relation disrupted by a breach of an international obligation is restored and appropriate reparation is made. The right to take countermeasures provides the injured state with a tool to secure the restoration of the legal relation and reparation of a breach unilaterally by its own means. Under the law of state responsibility, the lawful   taking   of   countermeasures   precludes   wrongfulness   of   otherwise

---

[714] 1992 UNFCCC, Art. 14(5).

[715] Treves 2009, p. 503.

[716] The KP Compliance Committee could be seen as an example of a collective enforcement mechanism.

[717] Kiss & Shelton 2007a, p. 81.

wrongful behaviour.[718] The use of countermeasures is justified by the previous wrongful conduct of the perpetrator state but is subject to strict conditions. First of all, as discussed in Chapter 3, countermeasures must be taken in a way that would allow the responsible state to comply with its obligations of cessation and reparation. Second, countermeasures are always temporary and must cease as soon as the responsible state has complied with its international obligations under the law of state responsibility. Third, they must be proportional to the injury sustained and finally, they may not belong to the list of prohibited countermeasures contained in ILC Article 50(1), such as the use of force or peremptory norms on international law.

A state injured by the deleterious effects of climate change could therefore resort to countermeasures to respond to another state's failure to take adequate mitigation measures. For example, an injured state could take trade-related environmental measures of the sort contained in the US legislative proposal described in Section 5.3.2 above. It could introduce an international reserve allowance requirement on imported goods or a similar border adjustment tax to offset carbon emissions of imported goods. As long as those countermeasures are in line with the law of state responsibility (see above), they need not be compatible with the WTO/GATT law.[719] All states are (potentially) affected by climate change and as such have the right to resort to countermeasures. It has happened in the past that states have taken action, including trade-related measures, against states perceived to be in breach of international obligations protecting a collective interest of a group of states or obligations owed to the international community as a whole. Examples include measures by the European Community Member States against the Federal Republic of Yugoslavia in 1998, which included the freezing of Yugoslav funds and an immediate flight ban; and suspension of landing rights of South African Airlines in the US in 1986 to encourage the South African government to adopt 'a non-racial democracy.'

It has been argued that obligations related to the mitigation of climate change are obligations *erga omnes* and states do not have to be individually injured to enforce those. At present, however, international law does not seem to recognize the right of states other than the injured state to take countermeasures to enforce a collective interest. Although there has been some state practice (involving a limited number of states) in support of such a right, the ILC did not deem it appropriate to include a provision on the right of states other than the injured state to take countermeasures.[720] In the climate change context, however, all states are injured states and therefore the taking of countermeasures by any state can be justified.

---

[718] See Section 3.2.2.4.1.

[719] Lefeber 2012, p. 348.

[720] 2001 Articles on Responsibility of States for Internationally Wrongful Acts, ILC Report on the work of its 53rd session, A/56/10, YILC, vol. II, Part Two, commentary to Art. 54, p. 355, para. 6.

5.6 Concluding Remarks

States are responsible for internationally wrongful acts they commit. If a state breaches its primary obligations under international law, it incurs international responsibility. In the present chapter, the relevant international obligations have been identified as obligations on the mitigation of climate change, obligations on climate change adaptation, and the customary obligation to prevent significant transboundary harm. Multilateral treaty-based obligations on climate change mitigation include quantified emissions reduction or limitation commitments and the adoption of mitigation policies and measures by developed countries. In contrast, developing states only have to formulate national or regional programmes containing measures to mitigate climate change but this obligation is contingent on the provision of financial resources and transfer of technology by industrialized countries. It has been concluded that, for practical reasons, it will not be possible to determine whether or not a developed state has complied with its KP reduction targets for the first commitment period before 2015 and that only industrialized states could potentially be held responsible for failure to adopt mitigation measures, which, given the absence of an international benchmark, would not be an easy task. With regard to obligations on climate change adaptation, it has been argued that developed states cannot be held accountable for not providing financial, technological, and capacity-building resources for adaptation efforts in developing countries.

It has been put forward that breach of the obligation to prevent significant transboundary harm can provide the sole legal basis for engaging the international responsibility of any state that fails to take adequate mitigation measures. Since states are duty-bound to regulate public and private conduct in areas subject to their jurisdiction or control in order to fulfil the due diligence requirement in their endeavours to prevent significant transboundary harm, they must adopt measures that would mitigate climate change by way of reduction or limitation of GHG emissions. It must be noted that the degree of due diligence that can legitimately be expected of developing countries must be lower than that to be exercised by industrialized states. If industrialized nations must take mitigation measures in accordance with the UNFCCC, it could be argued that developing states, too, must take some mitigation measures if they are to live up to their duty to prevent significant transboundary harm. It is in this context that the obligation to prevent significant transboundary harm may also provide a basis for challenging single major emission sources.

It has been argued that causation challenges may be difficult to overcome for the claimant state to obtain compensation for the climate change-related harm suffered. Meeting the *causa proxima* threshold required for reparation may present insurmountable challenges for the claimant state, particularly when harm has been caused by an extreme weather event. It has also been argued that retroactivity and the apportionment of responsibility may present additional difficulties.

According to the law of state responsibility, if in breach of the obligation to prevent significant transboundary harm, the perpetrator state is under an obligation to cease its wrongful behaviour and comply with the relevant procedural and substantive duties. In practice, it could enable the claimant state to seek injunctive relief as cessation amounts to the same. It has been submitted that in order to demonstrate breach of the primary norm, it is sufficient for the claimant state to show that GHG emissions originating in the territory within the respondent state's jurisdiction or control are merely capable of causing it significant transboundary harm. If a single emissions source in the respondent state's territory is challenged on the basis of the obligation to prevent significant transboundary harm, it is likewise enough to show that emissions from that source are capable of contributing to climate change, which can cause harm to the claimant state.

In order to comply with its obligation of cessation, the respondent state would have to start taking adequate mitigation measures required by the due diligence obligation to prevent significant transboundary harm from climate change. The adequacy of such measures would have to be determined in accordance with the principle of common but differentiated responsibilities and respective capabilities and would not be the same for all states.

In international law, the responsibility of the respondent state is normally invoked by the injured state. It has been argued that with respect to climate change, every state is an injured state. It has also been argued that international obligations on climate change mitigation are peremptory norms of international law and that no derogation from those obligations is permitted.

It has been concluded that the taking of adequate mitigation measures is necessary to comply with the duty to prevent significant transboundary harm. It has also been put forward that due to the voluntary nature of interstate dispute resolution, it may be difficult to find a competent court to hear a climate change-related claim. Therefore, in practice, if the responsible state is not taking adequate mitigation measures, unilateral action in the form of countermeasures may present an avenue of recourse for the injured state, which could take the form of trade-related environmental measures. Single emissions sources could also be challenged through diplomatic channels, as demonstrated by Micronesia's claim.

Conclusion

The idea of climate litigation at the interstate level is not new. Tuvalu considered suing certain industrialized states before the ICJ already in 2002. A small island developing nation, Tuvalu is extremely vulnerable to climate change and its very existence is threatened by the rising sea levels. In the absence of an effective international agreement on climate change, it is not surprising that litigation should come as an attractive option to states that are particularly affected by the injurious effects of climate change.

In general, interstate liability may serve a number of different purposes. First, it may perform a corrective function through the enforcement of primary international obligations *ex post facto*; second, it may dissuade states *ex ante facto* from engaging in conduct that would lead to liability thereby performing a preventive function; and third, liability may have a reparative function, which is aimed at shifting the injurious consequences of particular conduct from the victim to its author.[721] If the functions of liability are the 'whys' of litigation, then the focus of the present work can be said to have lain with the 'hows.' The present research has explored the pathways to interstate liability for climate change-related damage and concluded that litigation may be pursued under the law of state responsibility. Legal obstacles to state responsibility have been identified and ways to overcome them have been proposed.

The international political response to climate change being based on science, the first chapter of this treatise focused on the findings of the IPCC AR4. It addressed the causes and consequences of climate change; its impacts across the various geographical regions and sectors; and the main response strategies, namely mitigation and adaptation.

The second chapter described the international climate regime concluding that the current non-binding emissions reduction pledges together with the binding quantified emissions limitation and reduction commitments under the KP second commitment period would not be sufficient to keep the global average temperature rise under 2°C above preindustrial levels.[722] It has been put forward that the existence of a gap between the GHG emissions rise consistent with the current

---

[721] See Chapter 3.

[722] See 1992 UNFCCC, Art. 2; *Cancun Agreements: Outcome of the work of the* Ad Hoc *Working Group on Long-term Cooperative Action under the Convention*, UNFCCC Decision 1/CP.16 (2010), para. 4; *Outcome of the work of the* Ad Hoc *Working Group on Further Commitments for Annex I Parties under the Kyoto Protocol*, FCCC/KP/CMP/2012/L.9 (UNFCCC decision number not available at the time of writing), 8 December 2012.

pledges and that necessary to keep the global average temperatures under the 2°C threshold calls for adequate mitigation measures. It has been argued that liability can play a crucial role in ensuring that states take such measures.

In order to identify a legal framework for compelling states to take mitigation measures, the third chapter examined the different modalities of interstate liability used in international law, distinguishing between state responsibility and state liability. It has been found that the legal regime of state responsibility is engaged upon the commission of an internationally wrongful act, whereas state liability originates from the injurious consequences of lawful activities. The chapter discussed four conceptual approaches to state liability in international law giving rise to: (1) the obligation to pay compensation; (2) the obligation to negotiate a redress settlement; (3) the obligation to ensure prompt, adequate, and effective compensation; and (4) the obligation to take response action.

It has been determined that the absolute liability approach to interstate liability involving the obligation to pay compensation is unsuitable for addressing climate change-related damage for several reasons. First, since in international law the absolute liability approach is associated with ultrahazardous activities carrying significant risk, it cannot extend to the emission of GHGs because the emission of GHGs is not a new or dangerous activity in and of itself, and can only cause global warming when reaching high atmospheric concentrations. Second, states would be unlikely to assume absolute liability for climate change-related damage because most of GHG emissions originate from the private sector. Third, difficulties associated with tracing climate change-related damage in the territory of one state to GHG emissions originating in another cannot support the imposition of absolute liability on the source state, particularly given the fact that no state is carbon-neutral. It has been determined that although a liability mechanism imposing on states the duty to pay compensation would be an attractive option for countries suffering from the injurious consequences of climate change, in practice this approach is limited to space activities associated with ultrahazardous risks.

It has further been concluded that the approach to state liability based on the duty to negotiate on the reparation of transboundary harm when no internationally wrongful act has been committed is of little utility in the climate change context. This approach to liability has not taken hold in international law in general as it does not actually require states to reach an agreement, nor can it compel states to submit their dispute to a third party for settlement. Based on the occurrence of transboundary harm, the approach involving the obligation to negotiate a redress settlement would also entail complications with demonstrating the causal connection between the emissions of GHGs in one state and the harm suffered in another's territory (also see above).

Furthermore, it has been established that the approach to liability giving rise to the obligation to ensure prompt, adequate, and effective compensation cannot be relied upon in addressing climate change-related damage. While this approach would be attractive for climate change victims, such as members of coastal communities or inhabitants of low-lying islands, in international law its application has been limited to situations involving hazardous activities. It has been concluded that the emission of GHGs does not fall under the definition of 'hazardous,' and states cannot be required to ensure prompt, adequate, and effective compensation to victims of harm caused by activities that are considered non-hazardous. Another difficulty with applying this approach to climate change-related damage is that the obligation to ensure prompt, adequate, and effective compensation channels liability to operators. This could potentially result in unreasonably wide coverage as virtually every private party emits GHGs.

The last approach to state liability analysed in the third chapter has likewise been rejected as unsuitable for situations involving climate change-related damage. First, the obligation to take response measures is mainly directed at the *consequences* of harm to the environment and cannot be relied upon to ensure that states take measures to mitigate climate change. Second, while most of climate change-associated harm amounts to damage to property, the obligation to take response measures only extends to damage to the environment *per se*. Third, even if it could be argued that climate change is causing purely environmental damage, the obligation to take response measures would not be effective, particularly in cases of loss of territory to the sea level rise. Finally, since this approach engages operators' liability, additional problems would arise with identifying the operator responsible for the environmental harm caused by climate change.

Having concluded that none of the state liability schemes discussed can be successfully applied to climate change-related damage, Chapter 3 turned to the law of state responsibility in order to determine whether it can provide a legal framework for the accountability of states for climate change-related damage. While state liability pertains to situations when lawful conduct has resulted in harm, state responsibility is engaged in instances when a state has committed an internationally wrongful act. The chapter provided a thorough analysis of state responsibility by first focusing on its origins, including the notion of an internationally wrongful act, attributability of wrongful conduct to the responsible state, breach of an international obligation, retroactivity, and circumstances precluding wrongfulness; and distinguishing between due diligence obligations and absolute obligations, and obligations of conduct and obligations of result. Next, the chapter considered the content of state responsibility and the legal consequences of an internationally wrongful act, namely cessation and reparation (restitution, compensation, and satisfaction). Third, Chapter 3 dealt with breaches of obligations arising under peremptory norms of international law and, finally, the implementation of state responsibility was examined. It has been concluded that, as a general framework, the law of state responsibility has the potential to encompass accountability of states for climate change-related damage.

Given the absence of international jurisprudence on interstate liability for climate change-related damage, Chapter 4 examined a selection of legal actions launched in various national jurisdictions in order to identify the legal hurdles encountered by plaintiffs in the process of litigating climate change. It has been submitted that the relevance of domestic litigation for potential interstate claims can be justified through the notion of general principles of law. Legal challenges have been identified for three groups of climate change cases: (1) claims related to procedural injury; (2) claims for injunctive and/or declaratory relief; and (3) claims for compensation. They have included issues related to non-justiciability of political questions, standing, causation, attribution, and retroactivity. An analysis of climate change litigation in domestic courts has shown that albeit claims related to procedural injury have enjoyed the most degree of success, procedural justice cannot offer any immediate relief to climate change plaintiffs. Claims for compensation have thus far been unsuccessful and while suing for damages could be an attractive strategy for climate change claimants, actions for damages can only address adaptation to, and remediation of, climate change that is either happening or has already taken place. Conversely, injunctions are associated with mitigation action due to their preventive potential and as such can be effective in combating climate change. Injunctive relief claims have enjoyed some degree of success in domestic courts.

In order to evaluate whether the law of state responsibility can provide a legal framework for the accountability of states for climate change-related damage and whether it can provide an effective remedy in such situations, Chapter 5 undertook to identify primary obligations relating to climate change and analyse the practical application of the secondary rules of state responsibility in the international climate policy context. Basically, Chapter 5 has sought to apply the legal framework of the law of state responsibility to the climate change problem. Having distinguished the relevant primary obligations of states, namely international obligations on climate change mitigation; obligations on climate change adaptation; and the customary obligation to prevent significant transboundary harm, the fifth chapter determined the origins of state responsibility for breaches of those obligations by industrialized states party to the KP; EITs; industrialized states not party to the KP; and developing states.[723]

With respect to obligations on climate change adaptation, it has been concluded that industrialized countries' failure to finance adaptation measures in developing states or their failure to provide to developing countries technology and capacity-building resources for adaptation purposes cannot give rise to state responsibility.

With regard to obligations on climate change mitigation, it has been submitted that since the international climate regime only requires industrialized states to adopt mitigation policies and measures, and developing states are not legally bound to take such measures, only developed countries' responsibility can be engaged for failure to do so. It has been argued that in practice it would be difficult to demonstrate breach in the absence of an internationally accepted benchmark for mitigation policies and measures. As far as quantified GHG emissions reduction or limitation targets are concerned, it has been put forward that in order to launch a claim against an industrialized state for failure to meet its target, the claimant state would have to wait until after the obligation is due following the end of the first commitment period.

It has been submitted that, today, breach of the obligation to prevent significant transboundary harm can provide the sole legal basis for engaging the international responsibility of any state that fails to take adequate mitigation measures. Under the KP, industrialized countries are only required to achieve their quantified emissions reduction or limitation targets under the first commitment period by the end of 2012.[724] Under the KP review and reporting procedures, assessment of non-compliance will not occur before 2015. It has been argued that the claimant state could argue already now that if a developed state is on a clear trajectory towards non-compliance with its KP target, it cannot be considered to have acted with due diligence in its efforts to comply with the customary duty to prevent significant transboundary harm. Furthermore, since the obligation to prevent significant transboundary harm is part of customary international law, it does not differentiate between developed and developing states. Industrialized states, party or non-party to the KP alike, as well as developing states can be held responsible for breaching the obligation, *i.e.* lack of due diligence in a state's efforts to prevent significant transboundary harm would give rise to its international responsibility. It has been submitted that in order to prevent significant transboundary harm, states have a procedural obligation to regulate activities within their jurisdiction or control. Substantively, in order to prevent significant transboundary harm associated with

---

[723] See 1992 UNFCCC, Art. 3(1) (the principle of common but differentiated responsibilities and respective capabilities).

[724] Following the Doha amendment, under the second commitment period, industrialized countries are required to achieve their individual emissions limitation and reduction commitments by the end of 2020.

climate change, they must take mitigation action. It has been argued that such mitigation action must be compatible with the global temperature rise of less than 2°C above preindustrial levels. [725] It has further been argued that, since the obligation to prevent extends to actual as well as potential harm, the claimant state would have to show that GHG emissions originating in the respondent state are merely capable of causing damage by way of their contribution to global warming.

Chapter 5 has established the origins of state responsibility for industrialized states, including EITs, party to the KP (*e.g.* Canada); industrialized states not party to the KP (*e.g.* the US); and developing states (*e.g.* China). It has been submitted that despite Canada's withdrawal from the KP, it is responsible for breaching at least some of its obligations under the Protocol and could additionally be challenged for acting in breach of its customary duty to prevent significant transboundary harm. It has been argued that the US as well as heavily emitting developing states could likewise be held responsible under international law for breaching the obligation to prevent significant transboundary harm.

It has also been put forward that the law of state responsibility can provide a legal basis for seeking injunctive relief that would require the responsible state to take adequate mitigation measures or discontinue the construction or modernization of a single major GHG emitting source until the relevant procedural and substantive duties are fulfilled. It has been submitted that under the law of state responsibility, cessation and guarantees of non-repetition would in effect amount to injunctive relief. A declaration of wrongfulness by an international court calling for cessation of wrongful conduct in respect of a state failing to take adequate mitigation measures and thus acting in breach of the obligation to prevent significant transboundary harm would be equivalent to an injunction. It has also been suggested that obligations on climate change mitigation be considered as peremptory norms of international law, no derogation from which is permitted.

It has been submitted that since all states are affected by climate change and can therefore be considered injured, all states have the *locus standi* to invoke state responsibility. It has also been recognized that it may not be easy to submit a dispute regarding alleged breaches of obligations related to climate change mitigation or the obligation to prevent significant transboundary harm to a third party for settlement because the acceptance of international courts' jurisdiction by states is voluntary. It has therefore been suggested that in situations, when not all parties to the dispute have recognized the jurisdiction of a particular court, self-help, which could take the form of trade-related environmental countermeasures,

---

[725] *Cancun Agreements: Outcome of the work of the* Ad Hoc *Working Group on Long-term Cooperative Action under the Convention*, UNFCCC Decision 1/CP.16 (2010), para. 4.

may be an attractive option for a state suffering from the injurious consequences of climate change.

Now that the main findings of the present treatise have been summarized, it is important to draw some conclusions from the analysis of domestic litigation related to climate change (Chapter 4) in terms of what can be learnt from it for the purposes of interstate responsibility. The legal hurdles singled out in the analysis of domestic jurisprudence (political question doctrine, standing, causation, attribution, and retroactivity) do not all recur to the same extent in the context of interstate responsibility. Yet, the way domestic courts have treated these issues, *e.g.* causation, can provide helpful insights for the benefit of interstate claims that may be brought in the future.

The first legal obstacle identified is the political question doctrine. It has not been a major obstacle in procedural injury claims but presented considerable difficulties in claims for compensation and actions for injunctive relief. Some courts have ruled that claims before them presented questions of a political nature that should be left to the executive and legislative branches of government. The political question doctrine is based on the domestic governance model of the separation of powers. It has no analogue at the interstate level due to the fundamentally different nature of the international legal order whereby all states enjoy sovereign equality.

The second hurdle in domestic climate change litigation is standing. Generally, it has not been problematic in procedural injury claims but created substantial difficulties for plaintiffs seeking compensation or injunctive relief. In US case law, the requirements for standing have been identified as causation (see below), redressability, and injury, all of which need to be satisfied for a claim to proceed on the merits. Whereas the requirements of causation and redressability in domestic claims have been hard to meet, plaintiffs have had little difficulty showing injury.

At the interstate level, standing is also a requirement for bringing a claim. In order to have standing under the law of state responsibility, the claimant state has to demonstrate injury. It is unlikely to present major challenges because, as far as climate change is concerned, every state is an injured state and as such, has the *locus standi* under the law of state responsibility (see Section 5.5.1).

The third, and the most significant, obstacle climate change plaintiffs have grappled with is causation. An analysis of procedural injury claims has revealed that, in US case law, requirements for causation as an element of standing have been relaxed. It is important to mention that although the plaintiffs in *Massachusetts v. EPA* – one of the most prominent climate change cases in the US – sought an injunction, the injury complained of was of a procedural character. Although the plaintiffs also satisfied the traditional standing requirements, their success could be attributed, in part, to the relaxed causation standard involved in establishing a procedural right. Australian procedural injury claimants have largely been successful in establishing general as well as specific causation. Australian courts have accepted the general causal link between anthropogenic GHG emissions and climate change. They have also accepted specific causation by recognizing that contributions from a particular source of GHG emissions must be taken into account when making an EIA.

An interstate claimant alleging procedural injury would also appear to have a good chance for successfully demonstrating causation. In fact, Micronesia's challenge of a power plant modernization project in the Czech Republic, albeit submitted through diplomatic channels, suggests that it can be done. Micronesia was able to show that the transboundary impact of a single emissions source must be reflected in the EIA because of its contribution to climate change, which, through its manifestations, causes harm in its territory. Failure to exercise other procedural duties inherent in the obligation to prevent significant transboundary harm, such as the duty to notify a potentially affected state or failure to monitor transboundary impacts of a project, could also be challenged.

An examination of domestic jurisprudence has revealed a relaxed standard of causal proof in procedural injury cases. It could be argued that it is indicative of a general principle of law. Therefore, a procedural injury claim at the interstate level would likely not be hampered by causal ambiguity.

In domestic claims for injunctive relief, causation as part of the standing inquiry has been a key challenge for climate change plaintiffs, even though the 'fairly traceable' standard is significantly lower than the one involved in the *causa proxima* test used to decide claims on the merits. The courts have indicated that considerable evidentiary challenges must be expected at the merits stage. Although, as pointed out earlier, the successful outcome of *Massachusetts v. EPA* was, to some extent, rooted in the procedural character of the injury alleged, the plaintiffs were also able to meet the traditional standing requirements and show causation to the 'fairly traceable' degree.

In order to seek an injunction at the interstate level, it must be determined that a primary norm has been breached, which would give rise to the obligation of cessation. It has been argued that in proving breach of the primary obligation to prevent significant transboundary harm, causation hinges on the foreseeability of potential harm. In more specific terms, the claimant state would need to demonstrate that GHG emissions originating in the respondent state are capable of causing it foreseeable harm. Therefore, it has been argued, the respondent state must exercise due diligence and start taking adequate mitigation measures.

An evaluation of domestic claims for injunctive relief suggests that, should they be decided on the merits, causation would likely prove an insurmountable obstacle. At the interstate level, the *causa proxima* test would also likely be impossible to meet. However, the need for proving the causal link between the injury and the internationally wrongful act only arises in reparation claims, *e.g.* claims for compensation (see below). Therefore, injunctive relief sought on the basis of the obligation of cessation may provide the sole pathway to interstate liability for climate change-related damage.

A study of domestic claims for compensation has shown that plaintiffs have had enormous difficulties with proving causation as an element of standing. The courts have indicated that evidentiary challenges to demonstrating causation on the merits would likely be impossible to overcome.

It has been argued that, under the law of state responsibility, the *causa proxima* test involved in sustaining a claim for reparation would likely be impossible to meet. Domestic plaintiffs' lack of success, too, seems to point in the same direction.

The fourth legal challenge identified in the analysis of domestic jurisprudence is attribution. National courts have addressed the question of attribution of harm to a particular respondent as part of discussion on causation, demonstrating which has presented enormous challenges to plaintiffs. The courts have noted, however, that there is a lack of clarity as to whose GHG emissions have caused the plaintiffs' injuries.

Albeit related to attribution in domestic case law, attributability in the law of state responsibility is a distinctly different concept. Attributability of wrongful conduct to the state is one of the elements of an internationally wrongful act. It has not been contested that GHG emissions from private actors cannot be attributed to the state but it has been argued that failure to regulate the conduct of private actors can. Therefore, unlike in domestic case law, attributability of wrongful conduct is unlikely to become a hurdle to interstate liability.

The last legal issue presented in the analysis of domestic case law is that of retroactivity. Domestic courts have not explicitly dealt with it in the context of climate change. It has been argued, however, that it would be of relevance to compensation claims at the merits stage of the proceedings because of the fact that claims for compensation concern injuries that are actual; not potential harms. Cut-off dates have been used to address retroactivity in, for example, US tobacco cases. At the interstate level, retroactivity is likely to pose challenges in the context of reparation. It has been submitted that the introduction of a cut-off date would likely preclude the responsibility of the respondent state because climate change has been caused by anthropogenic GHG emissions released worldwide from the time of the Industrial Revolution.

The present treatise has assessed the possible forms of redress available to the claimant state suffering climate change-related injuries. Following a careful examination of legal challenges likely to be encountered by the claimant state, the conclusion is that the law of state responsibility can provide the legal framework for an interstate claim seeking injunctive relief on the basis of the obligation of cessation.

Bibliography

Abate, R.S. (2008). 'Massachusetts v. EPA and the future of environmental standing in climate change litigation and beyond.' 33 *Wm. & Mary Envtl. L. & Pol'y Rev.* 121

Aguilar, S., Appleton, A., Dafoe, J., Doran, P., Kosolapova, E., McColl, V., Mead, L. & Recio, E. (2011). Summary of the Durban Climate Change Conference. *Earth Negotiations Bulletin.* 28 November-9 December 2011. Available from: <www.iisd.ca/download/pdf/enb12534e.pdf>

Benzoni, F. (2008). 'Environmental standing: who determines the value of other life?', 18 *Duke Envtl. L. & Pol'y F.* 347

Birnie, P., Boyle, A. & Redgwell, C. (2009). *International Law & the Environment* (3rd ed.). New York: Oxford University Press

Böckstiegel, K. (1992). Case Law on Space Activities. In N. Jasentuliyana (Ed.), *Space Law: Development and Scope* (206-218). Westport: Praeger Publishers

Bodansky, D., Crook, J.R. & Crawford, J. (2002). 'The ILC's articles on responsibility of states for internationally wrongful acts: a retrospect.' 96 *AMJIL* 874

Boutrous, T.J., Jr. & Lanza, D. (2008). 'Global warming tort litigation: the real "public nuisance."' *Ecology Law Currents* 35

Brunnée, J. (2004). 'Of sense and sensibility: reflections on international liability regimes as tools for environmental protection.' 53 *Int'l & Comp. L.Q.* 351

Bryner, G. (2007). 'The rapid evolution of climate change law.' *Utah B.J.* 22

Cullet, P. (2007). 'Liability and redress for human-induced global warming: towards an international regime.' 43A *Stan. J. Int'l L.* 99

Daly, P. (2010). 'Justiciability and the 'political question' doctrine.' *P.L.* Jan. 160

De Sadeleer, N. (2010). The Principles of Prevention and Precaution in International Law: Two Heads of the Same Coin? In M., Fitzmaurice, D.M. Ong & P. Merkouris (Eds.), *Research Handbook on International Environmental Law* (182-199). Cheltenham: Edward Elgar

Desai, B.H. & Sidhu, B. (2012). 'International environmental dispute settlement.' 42/2 *Environmental Policy and Law*

Drumbl, M.A. (2006). Trail Smelter and the International Law Commission's Work on State Responsibility for Internationally Wrongful Acts and State Liability. In R.M. Bratspies & R.A. Miller (Eds.), *Transboundary Harm in International Law* (85-98). New York: Cambridge University Press

Durrant, N. (2007). 'Tortious liability for greenhouse gas emissions? Climate change, causation and public policy considerations.' 7 *QUT LJJ* 403

Eschen Mills, M. (2007). 'The global warming case: Massachusetts v. Environmental Protection Agency.' 7 *Sustainable Dev. L. & Pol'y* 67

Farber, D. (2008). 'Basic compensation for victims of climate change.' 38 *Envtl. L. Rep. News & Analysis* 10521

Faure, M. G., Nollkaemper, A. & Amsterdam International Law Clinic (2007). *Climate Change Litigation Cases.* Amsterdam: Milieudefensie

Faure, M. G. & Nollkaemper, A. (2007). 'International liability as an instrument to prevent and compensate for climate change.' 43A *Stan. J. Int'l L.* 123

Fitzmaurice, M. (2004). 'The Kyoto Protocol compliance regime and treaty law.' 8 *Singapore Yearbook of International Law* 23

Fitzmaurice, M. (2007). International Responsibility and Liability. In D. Bodansky, J. Brunnée & E. Hey (Eds.), *The Oxford Handbook of International Environmental Law* (1010-1035). New York: Oxford University Press

Fitzmaurice, M. (2010). 'Responsibility and Climate Change.' 53 *German Yearbook of International Law* 90

Fitzmaurice, M. & Redgwell, C. (2000). 'Environmental non-compliance procedures and international law,' 31 *Netherlands Yearbook of International Law* 35

Garnaut, R. (2008). Garnaut Climate Change Review. Available from: <www.garnautreview.org.au/2008-review.html>

Grossman, D.A. (2003). 'Warming up to a not-so-radical idea: tort-based climate change litigation.' 28 *Colum. Envtl. L.* 1

Gupta, J. (2007). 'Legal steps outside the Climate Convention: litigation as a tool to address climate change.' 16 *RECIEL* 76

Handl, G. (2006). Trail Smelter in Contemporary International Environmental Law: Its Relevance in the Nuclear Energy Context. In R.M. Bratspies & R.A. Miller (Eds.), *Transboundary Harm in International Law* (125-139). New York: Cambridge University Press

Hardman Reis, T. (2011). *Compensation for Environmental Damages under International Law: the Role of the International Judge.* Alphen aan den Rijn: Kluwer Law International

Heinzerling, L. (2008). 'Climate change in the Supreme Court.' 38 *Envtl. L.* 1

Hoss, C. (2005). State Responsibility, Liability and Environmental Protection. In R. Wolfrum, C. Langenfeld & P. Minnerop (Eds.), *Environmental Liability in*

International Law: Towards a Coherent Conception (455-494). Berlin: Erich Schmidt Verlag

Hsu, S.-L. (2008). 'A realistic evaluation of climate change litigation through the lens of a hypothetical lawsuit.' 79 *U. Colo. L. Rev.* 701

Johnson, M. (2006). 'Liability for environmental damage in Antarctica: the adoption of Annex VI to the Antarctic Environmental Protocol.' 19 *Geo. Int'l Envtl. L. Rev.* 33

Kiss, A. & Shelton, D. (2007a). *Guide to International Environmental Law.* Leiden: Martinus Nijhoff Publishers

Kiss, A. & Shelton, D. (2007b). Strict Liability in International Environmental Law. In T.M. Nndiaye & R. Wolfrum (Eds.), *Law of the Sea, Environmental Law and Settlement of Disputes. Liber Amicorum Judge Thomas A. Mensah.* (1131-1151). Leiden: Martinus Nijhoff Publishers

Klabbers, J. (2007). Compliance Procedures. In D. Bodansky, J. Brunnée & E. Hey (Eds.), *The Oxford Handbook of International Environmental Law* (995-1009). New York: Oxford University Press

Langenfeld, C. & Minnerop, P. (2005). Environmental Liability Provisions in International Law. In R. Wolfrum, C. Langenfeld & P. Minnerop (Eds.), *Environmental Liability in International Law: Towards a Coherent Conception* (3-158). Berlin: Erich Schmidt Verlag

Lean, J., Beer, J. & Bradley, R. (1995). 'Reconstruction of solar irradiance since 1610: implications for climate change.' 22 *Geophysical Research Letters* 3195

Lefeber, R. (1996). *Transboundary Environmental Interference and the Origin of State Liability.* The Hague: Kluwer Law International

Lefeber, R. (2009). *An Inconvenient Responsibility.* Utrecht: Eleven International Publishing

Lefeber, R. (2012). Climate Change and State Responsibility. In R. Rayfuse & S. Scott (Eds.), *International Law in the Era of Climate Change* (321-349). Cheltenham: Edward Elgar

Lefeber, R. (2012). 'The legal significance of the Nagoya-Kuala Lumpur Supplementary Protocol: the result of a paradigm evolution.' Amsterdam Law School Legal Studies Research Paper No. 2012-87. Centre for Environmental Law and Sustainability Research Paper No. 2012-02

Lefeber, R. & Oberthür, S. (2010). 'Holding countries to account: the Kyoto Protocol's compliance system revisited after four years of experience.' *Climate Law* 1

Loibl, G. (2010). Compliance Procedures and Mechanisms. In M. Fitzmaurice, D.M. Ong & P. Merkouris (Eds.), *Research Handbook on International Environmental Law* (426-449). Cheltenham: Edward Elgar

MacLeod, N. (2001). Extinction. In *Encyclopaedia of Life Sciences*. London: Macmillan. Available from:
<www.nhm.ac.uk/hosted_sites/paleonet/MacLeod/pdfs/extinction!.pdf>
Mank, B.C. (2005). 'Standing and global warming: is injury to all injury to none?' 35 *Envtl. L.* 1
Mank, B.C. (2009). 'Standing and future generations: does Massachusetts v. EPA open standing for generations to come?' 34 *Colum. J. Envtl. L.* 1
Murphy, J.F. (2010). *The Evolving Dimensions of International Law: Hard Choices for the World Community*. New York: Cambridge University Press

Okowa, P. (2000). *State Responsibility for Transboundary Air Pollution in International Law*. Oxford: Oxford University Press
Ong, D.M. (2010). International Legal Efforts to Address Human-Induced Global Climate Change. In M. Fitzmaurice, D.M. Ong & P. Merkouris (Eds.), *Research Handbook on International Environmental Law* (450-470). Cheltenham: Edward Elgar

Peel, J. (2007). 'The role of climate change litigation in Australia's response to global warming.' 24 *EPLJ* 90
Pidot, J.R. (2006). Global Warming in the Courts. An Overview of Current Litigation and Common Legal Issues. Georgetown Environmental Law & Policy Institute, Georgetown University Law Center. Retrieved on 6 April 2009 from:
<www.law.georgetown.edu/gelpi/current_research/documents/GWL_Report.pdf>
Pidot, J.R. (2007). Global Warming in the Courts: the *Massachusetts v EPA* Decision and its Implications. Georgetown Environmental Law & Policy Institute, Georgetown University Law Center. Retrieved on 14 April 2009 from:
<www.law.georgetown.edu/gelpi/current_research/documents/GWL_Update_4.24.07.pdf>
Posner, E.A. (2007). 'Climate change and international human rights litigation: a critical appraisal.' 155 *U. Pa. L. Rev* 1925
Preston, B. (2010). Climate change litigation. Paper delivered to Climate Change Governance after Copenhagen Conference, Hong Kong, 4 November 2010. Available from:
<www.lec.lawlink.nsw.gov.au/agdbasev7wr/_assets/lec/m4203011721754/preston_climate%20change%20litigation%20-%20a%20conspectus.pdf>

Quentin-Baxter, R.Q. (1982). Third report on international liability for injurious consequences arising out of acts not prohibited by international law. A/CN.4/360 & Corr.1, YILC, vol. II, Part One (1982)

Rose, A. (2007). 'Gray v Minister for Planning: the rising tide of climate change litigation in Australia.' *Syd. L. Rev.* 28

Sands, P. (2003). *Principles of International Environmental Law* (2nd ed.). Cambridge: Cambridge University Press

Schwarte, C. & Byrne, R. (2010). International Climate Change Litigation and the Negotiation Process. Foundation for International Environmental Law and Development working paper. Retrieved on 1 March 2010 from: <www.field. org.uk/files/FIELD_cclit_long_Oct.pdf>

Shibata, A. (2009). How to Design an International Liability Regime for Public Spaces: the Case of the Antarctic Environment. In T. Komori & K. Wellens (Eds.), *Public Interest Rules of International Law: towards Effective Implementation* (347-372). Farnham: Ashgate

Slade, T.N. (2007). 'Climate change: the human rights implications for small island developing countries.' 37 *Environmental Policy and Law* 215

Smith, J. & Shearman, D. (2006). *Climate Change Litigation: Analysing the Law, Scientific Evidence & Impacts on the Environment, Health & Property.* Adelaide: Presidian Legal Publications

Spier, J. (2012). *Shaping the Law for Global Crises: Thoughts about the Role the Law could Play to Come to Grips with the Major Challenges of Our Time.* Utrecht: Eleven International Publishing

Stern, N. (2006). Stern Review on the Economics of Climate Change. Available from: <http://webarchive.nationalarchives.gov.uk/+/http://www.hm-treasury.gov.uk/stern_review_report.htm>

Stevenson, D. (2007). 'Special solicitude for state standing: Massachusetts v. EPA.' 112 *Penn St. L. Rev.* 1

Sugar, M. (2007). 'Massachusetts v. Environmental Protection Agency.' 31 *Harv. Envtl. L. Rev.* 531

Tice, V. (2008). 'From Vermont's maples to Wybong's olives: cross-cultural lessons from climate change litigation in the United States and Australia.' 10 *Asian-Pac. L. & Pol'y J.* 292

Tol, R. & Verheyen, R. (2001). Liability and Compensation for Climate Change Damages – a Legal and Economic Assessment. Centre for Marine and Climate Research, Hamburg University. Working paper FNU-9. Retrieved on 1 March 2011 from: <www.fnu.zmaw.de/fileadmin/fnu-files/publication/working-papers/liability.pdf>

Tomuschat, C. (2011). Global Warming and State Responsibility. In H. Hestermeyer *et al.* (Eds.), *Law of the Sea in Dialogue* (3-29). Heidelberg: Springer

Treves, T. (2009). The Settlement of Disputes and Non-compliance Procedures. In T. Treves *et al.* (Eds.), *Non-compliance Procedures and Mechanisms and the Effectiveness of International Environmental Agreements* (499-518). The Hague: TMC Asser Press

Verheyen, R. (2005). *Climate Change Damage and International Law – Prevention Duties and State Responsibility.* Leiden: Martinus Nijhoff Publishers

Voigt, C. (2008). 'State responsibility for climate change damages.' 77 *Nordic Journal of International Law* 1

Weinberg, P. (2008). ''Political questions': an invasive species infecting the courts.' 19 *Duke Envtl. L. & Pol'y F.* 155

Yamin, F. & Depledge, J. (2004). *The International Climate Change Regime: a Guide to Rules, Institutions and Procedures.* Cambridge: Cambridge University Press

Table of Treaties and Other Selected Instruments

1. Treaties and European Directives

1959 Antarctic Treaty. UNTS, vol. 402, p. 71 (1961)

1960 OECD Convention on Third Party Liability in the Field of Nuclear Energy. UNTS, vol. 956, p. 263 (1974)

1962 Convention on the Liability of Operators of Nuclear Ships. 57 AJIL 268 (1963)

1963 IAEA Convention on Civil Liability for Nuclear Damage. UNTS, vol. 1063, p. 265 (1977)

1963 OECD Convention Supplementary to the Convention on Third Party Liability in the Field of Nuclear Energy. UNTS, vol. 1041, p. 358 (1977)

1967 Treaty on Principles Governing the Activities of States in the Exploration and Use of Outer Space, including the Moon and Other Celestial Bodies. United Nations Treaties and Principles on Outer Space. Text of treaties and principles governing the activities of States in the exploration and use of outer space, adopted by the United Nations General Assembly. ST/SPACE/11. United Nations, New York, 2002, p. 3

1969 Council of Europe Draft Convention on the Protection of Fresh Water against Pollution. CECA Doc. 2561 (1969)

1969 IMCO International Convention on Civil Liability for Oil Pollution. UNTS, vol. 973, p. 3 (1975)

1969 Vienna Convention on the Law of Treaties. UNTS, vol. 1155, p. 331 (1987)

1971 Convention Relating to Civil Liability in the Field of Maritime Carriage of Nuclear Material. UNTS, vol. 974, p. 255 (1975)

1971 Ramsar Convention on Wetlands of international importance especially as waterfowl habitat. UNTS, vol. 996, p. 245 (1976)

1972 Convention on International Liability for Damage Caused by Space Objects. United Nations Treaties and Principles on Outer Space. Text of treaties and principles governing the activities of States in the exploration and use of outer space, adopted by the United Nations General Assembly. ST/SPACE/11. United Nations, New York, 2002, p. 13

1974 Exchange of Notes Between the United States of America and Canada Constituting an Agreement Relating to Liability for Loss and Damage from Certain Rocket Launches. UNTS, vol. 992, p. 97 (1975)

1979 Geneva Convention on Long-Range Transboundary Air Pollution. UNTS, vol. 1302, p. 217 (1992)

1982 UN Convention on the Law of the Sea. UNTS, vol. 1833, p. 3 (1998)

1985 ASEAN Agreement on the Conservation of Nature and Natural Resources. 15 *Environmental Policy and Law* 64 (1985)

1985 Helsinki Protocol to the 1979 Convention on Long-Range Transboundary Air Pollution on the Reduction of Sulphur Emissions or their Transboundary Fluxes by at least 30 per cent. UNTS, vol. 1480, p. 215 (1997)

1987 CMEA Convention on Liability for Damage Caused by Radiological Accidents in International Carriage of Irradiated Nuclear Fuel from Nuclear Power Plants. Unpublished, for references see International Liability for Injurious Consequences Arising out of Acts not Prohibited by International Law (Prevention of Transboundary Damage from Hazardous Activities). Doc. A/CN.4/501, YILC, vol. II, Part One (1999); see also Lefeber (1996)

1987 Montreal Protocol on Substances that Deplete the Ozone Layer. UNTS, vol. 1522, p. 3 (1997)

1988 Sofia Protocol to the 1979 Convention on Long-Range Transboundary Air Pollution Concerning the Control of Emissions of Nitrogen Oxides or their Transboundary Fluxes. UNTS, vol. 1593, p. 287 (1998)

1989 UNECE Convention on Civil Liability for Damage Caused During Carriage of Dangerous Goods by Road, Rail and Inland Navigation Vessels. Doc. ECE/TRANS/79 (1989)

1991 Protocol on Environmental Protection to the 1959 Antarctic Treaty. 30 ILM 1455 (1991)

1992 Convention for the Protection of the Marine Environment of the North-East Atlantic. UNTS, vol. 2354, p. 67 (2006)

1992 Convention on Biological Diversity. UNTS, vol. 1760, p. 79 (2001)

1992 UNECE Convention on the Protection and Use of Transboundary Watercourses and International Lakes. 31 ILM 1312 (1992)

1992 UNECE Convention on the Transboundary Effects of Industrial Accidents. 31 ILM 1330 (1992)

1992 UN Framework Convention on Climate Change. UNTS, vol. 1771, p. 107 (2000)

1993 Convention on Civil Liability for Damage Resulting from Activities Dangerous to the Environment (Lugano Convention). Available from: <http://conventions.coe.int/Treaty/en/Treaties/Html/150.htm>

1994 UN Convention to Combat Desertification. UNTS, vol. 1954, p. 3 (1999)

1997 IAEA Convention on Supplementary Compensation for Nuclear Damage. INFCIRC/567 (22 July 1998)

1997 Convention on the Law of the Non-navigational Uses of International Watercourses. Doc. A/51/869 (11 April 1997)

1997 Kyoto Protocol to the United Nations Framework Convention on Climate Change. UNTS, vol. 2303, p. 148 (2005)

1999 Basel Protocol on Liability and Compensation for Damage Resulting from the Transboundary Movement of Hazardous Wastes and their Disposal. Doc. UNEP/CHW.1/WG/1/9/2 (1999)

2003 Kiev Protocol on Civil Liability and Compensation for Damage Caused by the Transboundary Effects of Industrial Accidents on Transboundary Waters. Doc. ECE/MP.WAT/11-ECE/CP.TEIA/9 (2003)

Directive 2004/35/EC of the European Parliament and of the Council of 21 April 2004 on environmental liability with regard to the prevention and remedying of environmental damage. OJ 2004 L 143/56

2005 Annex VI to the Protocol on Environmental Protection to the Antarctic Treaty (Liability Arising from Environmental Emergencies). Available from: <www.ats.aq/documents/recatt/Att249_e.pdf>

Directive 2008/1/EC of the European Parliament and of the Council of 15 January 2008 concerning integrated pollution prevention and control. OJ 2008 L 24/8

2010 Nagoya - Kuala Lumpur Supplementary Protocol on Liability and Redress to the Cartagena Protocol on Biosafety. UNEP/CBD/BS/COP-MOP/5/17 (15 October 2010)

2. UN and Other Intergovernmental Documents

Stockholm Declaration on the Human Environment (16 June 1972). Available from: <www.unep.org/Documents.Multilingual/Default.asp?DocumentID=97&ArticleID =1503&l=en>

Draft Articles on Responsibility of States for Internationally Wrongful Acts. ILC Report on the work of its 28th session. A/31/10, YILC, vol. II, Part Two (1976)

Protection of global climate for present and future generations of mankind. GA Res. 43/53 (6 December 1988)

IPCC, First Assessment Report. Climate Change: The IPCC Scientific Assessment (1990)

SC Res. 687 (8 April 1991)

UNCC Governing Council Decision 7. S/AC.26/1991/7/Rev.1 (17 March 1992)

Rio Declaration on Environment and Development. Rio Declaration on Environment and Development. A/CONF.151/26 (Vol. I) (12 August 1992)

Statement of Principles for the Sustainable Management of Forests. A/CONF.151/26 (Vol. III) (14 August 1992)

Agenda 21. Environment and Development Agenda (1992). Available from: <www.unep.org/Documents.Multilingual/Default.asp?documentid=52>

Principles Relevant to the Use of Nuclear Power Sources in Outer Space. UN Res. 47/68 (1992)

Draft Articles on the Law of the Non-Navigational Uses of International Watercourses. ILC Report on the work of its 46th session. A/49/10, YILC, vol. II, Part Two (1994)

Principles Governing IPCC Work (1998), as amended in 2003, 2006 & 2012. Available from: <www.ipcc.ch/pdf/ipcc-principles/ipcc-principles.pdf>

ILC Report on the work of its 50th session. A/53/10, YILC, vol. II, Part Two (1998)

UNCC Governing Council Decision 132. S/AC.26/Dec.132 (21 June 2001)

Report and Recommendations Made by the Panel of Commissioners Concerning the First Instalment of 'F4' Claims. S/AC.26/2001/16 (22 June 2001)

Articles on Responsibility of States for Internationally Wrongful Acts. ILC Report on the work of its 53rd session. A/56/10, YILC, vol. II, Part Two (2001)

Draft Articles on Prevention of Transboundary Harm from Hazardous Activities. ILC Report on the work of its 53rd session. A/56/10, YILC, vol. II, Part Two (2001)

Responsibility of States for internationally wrongful acts. GA Res. 56/83 (28 January 2002)

Report and Recommendations Made by the Panel of Commissioners Concerning the Second Instalment of 'F4' Claims. S/AC.26/2002/26 (3 October 2002)

UNCC Governing Council Decision 171. S/AC.26/Dec.171 (3 October 2002)

Report and Recommendations Made by the Panel of Commissioners Concerning the Third Instalment of 'F4' Claims. S/AC.26/2003/31 (18 December 2003)

UNCC Governing Council Decision 212. S/AC.26/Dec.212 (18 December 2003)

Report and Recommendations Made by the Panel of Commissioners Concerning Part One of the Fourth Instalment of 'F4' Claims. S/AC.26/2004/16 (9 December 2004)

Report and Recommendations Made by the Panel of Commissioners Concerning Part Two of the Fourth Instalment of 'F4' Claims. S/AC.26/2004/17 (9 December 2004)

UNCC Governing Council Decision 234. S/AC.26/Dec.234 (9 December 2004)

UNCC Governing Council Decision 235. S/AC.26/Dec.235 (9 December 2004)

Report and Recommendations Made by the Panel of Commissioners Concerning the Fifth Instalment of 'F4' Claims. S/AC.26/2005/10 (30 June 2005)

UNCC Governing Council Decision 248. S/AC.26/Dec.248 (30 June 2005)

Allocation of loss in the case of transboundary harm arising out of hazardous activities. GA Res. 61/36 (4 December 2006)

Articles on Diplomatic Protection. ILC Report on the work of its 58th session. A/61/10, forthcoming in YILC (2006)

Principles on the Allocation of Loss in the Case of Transboundary Harm Arising out of Hazardous Activities. ILC Report on the work of its 58th session. A/61/10, forthcoming in YILC (2006)

Conclusions of the work of the Study Group on the Fragmentation of International Law: Difficulties arising from the Diversification and Expansion of International Law. ILC Report on the work of its 58th session. A/61/10, forthcoming in YILC (2006)

IPCC, Fourth Assessment Report. Climate Change 2007: Synthesis Report (2007)

IPCC, Fourth Assessment Report. Working Group I Report. The Physical Science Basis (2007)

IPCC, Fourth Assessment Report. Working Group II Report. Impacts, Adaptation and Vulnerability (2007)

IPCC, Fourth Assessment Report. Working Group III Report. Mitigation of Climate Change (2007)

Diplomatic Protection. GA Res. 62/67 (8 January 2008)

UNEP Guidelines for the Development of Domestic Legislation on Liability, Response Action and Compensation for Damage Caused by Activities Dangerous to the Environment. Adopted by the UNEP Governing Council in decision SS.XI/5, part B (26 February 2010)

Draft Articles on the Law of Transboundary Aquifers. ILC Report on the work of its 60th session. A/63/10, forthcoming in YILC (2008)

Climate change and its possible security implications. GA Res. 63/281 (11 June 2009)

Guidance for the development of a ship energy efficiency management plan (SEEMP). IMO Marine Environment Protection Committee, MEPC.1/Circ.683 (17 August 2009)

Guidelines for voluntary use of the ship energy efficiency operational indicator (EEOI). IMO Marine Environment Protection Committee, MEPC.1/Circ.684 (17 August 2009)

Interim guidelines for voluntary verification of the energy efficiency design index. IMO Marine Environment Protection Committee, MEPC.1/Circ.682 (17 August 2009)

Interim guidelines on the method of calculation of the energy efficiency design index for new ships. IMO Marine Environment Protection Committee, MEPC.1/Circ.681 (17 August 2009)

Climate change and its possible security implications. Report of the Secretary General GA Doc.6 4/350 (11 September 2009)

Declaration of the Climate Vulnerable Forum. Malé, Maldives (10 November 2009). Available from: <http://daraint.org/wp-content/uploads/2010/12/Declaration-of-the-CVF-FINAL2.pdf>

Consolidated statement of continuing ICAO policies and practices related to environmental protection – Climate change. ICAO Assembly Resolution A37-19 (2010)

Consolidated statement of continuing ICAO policies and practices related to environmental protection – General provisions, noise and local air quality. ICAO Assembly Resolution A37-18 (2010)

UNEP Emissions Gap Report (2010). Available from: <www.unep.org/publications/ebooks/emissionsgapreport/pdfs/GAP_REPORT_SUNDAY_SINGLES_LOWRES.pdf>

ILC Draft Articles on the Responsibility of International Organisations. ILC Report on the work of its 63rd session. A/66/10, forthcoming in YILC (2011)

The Future We Want: Zero-draft of the outcome document. Rio +20 – UNCSD (10 January 2012). Available from: <www.uncsd2012.org/futurewewant.html>

IPCC, Special Report. Managing the Risks of Extreme Events and Disasters to Advance Climate Change Adaptation (SREX) (2012)

3. MEA Decisions and Other Documents

Global Climate Change. UNEP Governing Council Decision 14/20 (17 June 1987)

WMO Executive Council Resolution 4 (1988)

Berlin Mandate: Review of the adequacy of Article 4, paragraph 2(a) and (b), of the Convention, including proposals related to a protocol and decisions on follow-up. UNFCCC Decision 1/CP.1 (1995)

Consideration of commitments for subsequent periods for Parties included in Annex I to the Convention under Article 3, paragraph 9. UNFCCC Decision 1/CMP.1 (2005)

Dialogue on long-term cooperative action to address climate change by enhancing implementation of the Convention. UNFCCC Decision 1/CP.11 (2005)

Guidelines for national systems under Article 5, paragraph 1, of the Kyoto Protocol. UNFCCC Decision 19/CMP.1 (2005)

Guidelines for review under Article 8 of the Kyoto Protocol. UNFCCC Decision 22/CMP.1 (2005)

Guidelines for the implementation of Article 6 of the Kyoto Protocol. UNFCCC Decision 9/CMP.1 (2005)

Guidelines for the preparation of the information required under Article 7 of the Kyoto Protocol. UNFCCC Decision 15/CMP.1 (2005)

Good practice guidance and adjustments under Article 5, paragraph 2, of the Kyoto Protocol. UNFCCC Decision 20/CMP.1 (2005)

Issues relating to adjustments under Article 5, paragraph 2, of the Kyoto Protocol. UNFCCC Decision 21/CMP.1 (2005)

Issues relating to the implementation of Article 8 of the Kyoto Protocol – 1. UNFCCC Decision 24/CMP.1 (2005)

Issues relating to the implementation of Article 8 of the Kyoto Protocol – 2. UNFCCC Decision 25/CMP.1 (2005)

Modalities and procedures for a clean development mechanism as defined in Article 12 of the Kyoto Protocol. UNFCCC Decision 3/CMP.1 (2005)

Modalities for the accounting of assigned amounts under Article 7, paragraph 4, of the Kyoto Protocol. UNFCCC Decision 13/CMP.1 (2005)

Modalities, rules and guidelines for emissions trading under Article 17 of the Kyoto Protocol. UNFCCC Decision 11/CMP.1 (2005)

Procedures and mechanisms relating to compliance under the Kyoto Protocol. UNFCCC Decision 27/CMP.1 (2005)

Standard electronic format for reporting Kyoto Protocol units. UNFCCC Decision 14/CMP.1 (2005)

Terms of service for lead reviewers. UNFCCC Decision 23/CMP.1 (2005)

US Climate Action Report 2010. Fifth National Communication on Climate Change of the United States of America under the UNFCCC. Available from: <http://unfccc.int/resource/docs/natc/usa_nc5.pdf >

Bali Action Plan. UNFCCC Decision 1/CP.13 (2007)

The 10-year strategic plan and framework to enhance the implementation of the Convention (2008-2018), UNCCD Decision 3/COP.8, ICCD/COP(8)/16/Add.1 (2007)

Question of Implementation – Greece. Kyoto Protocol Compliance Committee. Final Decision. CC-2007-1-8/Greece/EB (17 April 2008)

Question of Implementation – Canada. Kyoto Protocol Compliance Committee. Final Decision. CC-2008-1-6/Canada/EB (15 June 2008)

Question of Implementation – Croatia. Kyoto Protocol Compliance Committee. Final Decision. CC-2009-1-8/Croatia/EB (26 November 2009)

Copenhagen Accord. UNFCCC Decision 2/CP.15 (2009)

Ideas and proposals on paragraph 1 of the Bali Action Plan. Revised note by the Chair. FCCC/AWGLCA/2008/16/Rev.1 (15 January 2009)

Letter including autonomous mitigation actions from the Department of Climate Change, National Development and Reform Commission of China, to the UNFCCC Secretariat of 28 January 2010. Available from: <http://unfccc.int/files/meetings/cop_15/copenhagen_accord/application/pdf/chinacphaccord_app2.pdf>

Canada's Fifth National Communication on Climate Change. Submitted to the UNFCCC Secretariat on 12 February 2010. Available from: <http://unfccc.int/resource/docs/natc/can_nc5.pdf>

Question of Implementation – Bulgaria. Kyoto Protocol Compliance Committee. Final Decision. CC-2010-1-8/Bulgaria/EB (28 June 2010)

Climate Change Assessments. Review of the processes and procedures of the IPCC. InterAcademy Council (October 2010)

National Greenhouse Gas Inventory Data for the Period 1990-2008. Note by the secretariat. FCCC/SBI/2010/18 (4 November 2010)

Cancun Agreements: Outcome of the work of the *Ad Hoc* Working Group on Further Commitments for Annex I Parties under the Kyoto Protocol at its fifteenth session. UNFCCC Decision 1/CMP.6 (2010)

Cancun Agreements: Outcome of the work of the *Ad Hoc* Working Group on Long-term Cooperative Action under the Convention. UNFCCC Decision 1/CP.16 (2010)

Views on the work programme to consider approaches to address loss and damage. Submission of Grenada on behalf of AOSIS. FCCC/SBI/2011/MISC.1 (28 February 2011)

Question of Implementation – Romania. Kyoto Protocol Compliance Committee. Final Decision. CC-2011-1-8/Romania/EB (27 August 2011)

Question of Implementation – Ukraine. Kyoto Protocol Compliance Committee. Final Decision. CC-2011-2-9/Ukraine/EB (12 October 2011)

Question of Implementation – Lithuania. Kyoto Protocol Compliance Committee. Final Decision. CC-2011-3-8/Lithuania/EB (21 December 2011)

Annual report of the administrator of the international transaction log under the Kyoto Protocol. Note by the secretariat. FCCC/KP/CMP/2011/7 (2011)

Compilation of economy-wide emission reduction targets to be implemented by Parties included in Annex I to the Convention. Revised note by the secretariat. FCCC/SB/2011/INF.1/Rev.1 (2011)

Compilation of information on nationally appropriate mitigation actions to be implemented by Parties not included in Annex I to the Convention. Note by the secretariat. FCCC/AWGLCA/2011/INF.1 (2011)

Emissions trading and the project-based mechanisms. UNFCCC Decision 3/CMP.7 (2011)

Establishment of an *Ad Hoc* Working Group on the Durban Platform for Enhanced Action. UNFCCC Decision 1/CP.17 (2011)

Launching the Green Climate Fund. UNFCCC Decision 3/CP.17 (2011)

Modalities and procedures for carbon dioxide capture and storage in geological formations as clean development mechanism project activities. UNFCCC Decision 10/CMP.7 (2011)

Outcome of the work of the *Ad Hoc* Working Group on Further Commitments for Annex I Parties under the Kyoto Protocol at its sixteenth session. UNFCCC Decision 1/CMP.7 (2011)

Work programme on loss and damage. UNFCCC Decision 7/CP.17 (2011)

Canada's Climate Change Mitigation Plan. Presentation at workshop on clarification of the developed country parties' quantified economy-wide emission reduction targets and related assumptions and conditions, Bonn (17 May 2012). Available from: <http://unfccc.int/files/bodies/awg-lca/application/pdf/20120517_canada_1749.pdf>

Question of Implementation – Slovakia. Kyoto Protocol Compliance Committee. Preliminary finding, CC-2012-1-7/Slovakia/EB (14 July 2012)

Question of Implementation – Slovakia. Kyoto Protocol Compliance Committee. Final Decision. CC-2012-1-9/Slovakia/EB (17 August 2012)

Annual report of the Compliance Committee to the Conference of the Parties serving as the meeting of the Parties to the Kyoto Protocol. FCCC/KP/CMP/2012/6 (8 November 2012)

Outcome of the work of the *Ad Hoc* Working Group on Further Commitments for Annex I Parties under the Kyoto Protocol. FCCC/KP/CMP/2012/L.9, UNFCCC decision number not available at the time of writing (8 December 2012)

Approaches to address loss and damage associated with climate change impacts in developing countries that are particularly vulnerable to the adverse effects of climate change to enhance adaptive capacity. FCCC/CP/2012/L.4/Rev.1, UNFCCC decision number not available at the time of writing (8 December 2012)

Table of Cases

1. Permanent Court of International Justice

German Settlers in Poland, Advisory Opinion, 1923 PCIJ (Ser. B) No. 6
Factory at Chorzów (Germany v. Poland), Merits, Judgment 1928 PCIJ (Ser. A) No. 17

2. International Court of Justice

Corfu Channel (United Kingdom of Great Britain and Northern Ireland v. Albania), Judgment of 9 April 1949, ICJ Rep. 4
Case Concerning the Barcelona Traction, Light and Power Company, Limited (Belgium v. Spain), Judgment of 5 February 1970, ICJ Rep. 3
Case Concerning United States Diplomatic and Consular Staff in Tehran (United States of America v. Iran), Judgment of 24 May 1980, 1980 ICJ Rep. 3
Military and Paramilitary Activities In and Against Nicaragua (Nicaragua v. United States of America), Judgment of 27 June 1986,1986 ICJ Rep. 14
Legality of the Threat or Use of Nuclear Weapons, Advisory Opinion of 8 July 1996, 1996 ICJ Rep. 226
Case Concerning the Gabčíkovo-Nagymaros Project (Hungary v. Slovakia), Judgment of 25 September 1997, 1997 ICJ Rep. 7
Case Concerning Pulp Mills on the River Uruguay (Argentina v. Uruguay), Judgment of 20 April 2010, 2010 ICJ Rep. 14

3. International Tribunal for the Law of the Sea

Responsibilities and Obligations of States with Respect to Activities in the Area, Advisory Opinion of 1 February 2011, 2011 ITLOS Rep. 10

4. Arbitral Awards

British Claims in the Spanish Zone of Morocco (Spain *v.* United Kingdom), 1 May 1925, UNRIAA, Vol. II, p. 615 (2006)
Trail Smelter case (United States *v.* Canada), 16 April 1938 and 11 March 1941, UNRIAA, vol. III, p. 1905 (2006)

Rainbow Warrior (New Zealand *v.* France), 30 April 1990, UNRIAA, vol. XX, p. 215 (2006)

5. *Inter-American Commission on Human Rights*

Petition to the Inter American Commission on Human Rights Seeking Relief from Violations Resulting from Global Warming caused by Acts and Omissions of the United States, submitted by Sheila Watt-Cloutier, 7 December 2005

6. *National Decisions*

6.1 Australia

Aldous v. Greater Taree City Council [2009] NSWLEC 17
Australian Conservation Foundation v. Commonwealth [1980] 146 CLR 493
Australian Conservation Foundation v. Minister for Planning [2004] VCAT 2029
Charles & Howard Pty Ltd v. Redland Shire Council [2007] QCA 200
Gippsland Coastal Board v. South Gippsland SC & Ors (No 2) (includes Summary) (Red Dot) [2008] VCAT 1545
Gray v. Minister for Planning and Ors [2006] NSWLEC 720
Greenpeace Australia Ltd v. Redbank Power Company Pty Ltd [1994] NSWLEC 178
Minister for Planning v. Walker [2008] NSWCA 224
Northcape Properties Pty Ltd v. District Council of Yorke Peninsula [2008] SASC 57
Queensland Conservation Council Inc. v. Xstrata Coal Queensland P/L & Ors [2007] QCA 338
Re Xstrata Coal Queensland Pty Ltd & Ors [2007] QLRT 33
Taralga Landscape Guardians Inc. v. Minister for Planning and RES Southern Cross Pty Ltd [2007] NSWLEC 59
Thornton v. Adelaide Hills Council [2006] 151 LGERA 1
Walker v. Minister for Planning [2007] NSWLEC 741
Wildlife Preservation Society of Queensland Proserpine/Whitsunday Branch Inc. v. Minister for the Environment & Heritage & Ors [2006] FCA 736

6.3 Canada

Friends of the Earth – Les Ami(e)s de la Terre v. The Governor in Council and The Minister of the Environment, 2008 FC 1183, 20 October 2008

6.4 Germany

Bund für Umwelt- und Naturschutz Deutschland e.V. (BUND) & Germanwatch e.V. gegen Bundesminister für Wirtschaft und Arbeit, Verwaltungsgericht Berlin, Beschluss VG 10 A 215.04, 2006; unofficial English translation available from: <www.climatelaw.org/cases/case-documents/germany/de-export-jan06-eng.doc>

6.5 New Zealand

Genesis Power Ltd v. Franklin District Council [2005] NZRMA 541
Greenpeace v. Northland Regional Council & Mighty River Power Ltd [2006] NZHC 1212
Environmental Defence Society (Inc) v. Auckland Regional Council [2002] NZRMA 492
Meridian Energy Limited & Ors v. Wellington City Council and Wellington Regional Council [2007] W031/2007
Unison Networks Ltd v. Hastings District Council [2007] NZHC 1435

6.6 Nigeria

Jonah Gbemre v. Shell Petroleum Development Company Nigeria Ltd., Suit No. FHC/B/CS/53/05, Federal High Court of Nigeria, Benin Judicial Division, 2005

6.7 United Kingdom

Dimmock v. Secretary of State for Education and Skills [2007] EWHC 2288 (Admin)

6.8 United States

American Electric Power Company, Inc. v. Connecticut, 131 S.Ct. 2527, U.S., 2011
Baker v. Carr, 369 U.S. 186 (1962)
Black Mesa Water Coalition v. Salazar, 2012 WL 2848437 (D.Ariz. 2012)
Border Power Plant Working Group v. Department of Energy, 260 F.Supp. 2d 997 (S.D.Cal 2003)

California v. General Motors Corp., 2007 WL 2726871 (N.D.Cal.)

Center for Biological Diversity v. Abraham, 218 F.Supp. 2d 1143 (N.D.Cal. 2002)

Center for Biological Diversity v. Brennan, 571 F.Supp. 2d 1105 (N.D.Cal. 2007)

Center for Biological Diversity v. United States Department of the Interior, 563 F. 3d 466 (D.C.Circ. 2009)

Coalition for a Sustainable Future in Yucaipa v. City of Yucaipa, 198 Cal.App.4th 939 (Cal.App. 4 Dist. 2011)

Comer v. Nationwide Mutual Insurance Co., WL 1066645 (S.D.Miss.) (23 February 2006)

Comer v. Murphy Oil USA, Inc., 2007 WL 6942285 (S.D.Miss.) (20 August 2007) (NO. CIVA 1:05-CV436LGRHW)

Comer v. Murphy Oil USA, Inc., 585 F.3d 855, C.A.5 (Miss.), 2009

Comer v. Murphy Oil USA, Inc., Case 1:11-cv-00220-LG-RHW (S.D.Miss.) (20 March 2012)

Connecticut v. American Electric Power Co., Inc., 406 F.Supp. 2d 265 (S.D.N.Y. 2005)

Connecticut v. American Electric Power Co., Inc., 582 F.3d 309, C.A.2 (N.Y.), 2009

Engle v. Liggett Group, Inc., 945 So.2d 1246 (Fla. 2006)

Friends of the Earth, Inc. v. Laidlaw Environmental Services, Inc., 120 S.Ct. 693 (U.S. 2000)

Friends of the Earth, Inc. v. Mosbacher, 488 F.Supp. 2d 889 (N.D.Cal. 2007)

Friends of the Earth, Inc. v. Watson, 35 ELR 20179 (N.D.Cal. 2005)

Habitat and Watershed Caretakers v. City of Santa Cruz, 149 Cal.Rptr.3d 574 (Cal.App. 6 Dist. 2012)

Healdsburg Citizens for Sustainable Solutions v. City of Healdsburg, 206 Cal.App.4th 988 (Cal.App. 1 Dist. 2012)

Korsinsky v. EPA, 2005 WL 2414744 (S.D.N.Y.)

Lujan v. Defenders of Wildlife, 112 S.Ct. 2130 (U.S.Minn. 1992)

Marbury v. Madison, 5 U.S. (1 Cranch) 137 (1803)

Massachusetts v. EPA, 127 S.Ct. 1438, U.S., 2007

Native Village of Kivalina v. ExxonMobil Corp., 663 F.Supp.2d 863 (N.D.Cal. 2009)

Native Village of Kivalina v. ExxonMobil Corp., 2012 WL 4215921 (9th Cir.(Cal.)) (21 September 2012) (NO. 09-17490)

Natural Resources Defence Council v. Kempthorne, 506 F.Supp. 2d 322 (E.D.Cal. 2007)

Northwest Environmental Defense Center v. Owens Corning Corp., 434 F.Supp. 2d 957 (D.Or. 2006)

Miscellaneous Sources

American Clean Energy and Security Act 2009, H.R. 2454 (111th Congress). Available from: <http://democrats.energycommerce.house.gov/Press_111/20090701/hr2454_house.pdf>

Climate Case Chart. Columbia Center for Climate Change Law. Available from: <http://climatecasechart.com/>

Climate Change Plan for the Purposes of the Kyoto Protocol Implementation Act – 2007. Environment Canada Inquiry Centre (2007). Available from: <www.ec.gc.ca/doc/ed-es/p_123/CC_Plan_2007_e.pdf>

Government of Canada news release of 12 December 2007. Available from: <www.ec.gc.ca/default.asp?lang=En&n=714D9AAE-1&news=0F208D84-395E-4E78-8E6F-CCB906C30F5B>

Ministry of the Environment of the Czech Republic press release of 1 February 2010. Available from: <www.mzp.cz/en/news_100126>

Ministry of the Environment of the Czech Republic press release of 30 April 2010. Available from: <www.mzp.cz/en/news_100430_statement_Prunerov>

Ministry of the Environment of the Czech Republic press release of 20 October 2010. Available from: <www.mzp.cz/en/news_101020_Prunerov>

Palau seeks UN World Court opinion on damage caused by greenhouse gases. UN News Centre, 22 September 2011. Available from: <www.un.org/apps/news/story.asp?NewsID=39710&Cr=pacific+island&Cr1=>

Settlement Agreement Export-Import Bank of the United States of 6 February 2009. Available from: <www.climatelaw.org/cases/case-documents/us/exim.pdf>

Turning the Corner: An Action Plan to Reduce Greenhouse Gases and Air Pollution. Government of Canada news release of 26 April 2007. Available from: <www.ec.gc.ca/default.asp?lang=En&n=714D9AAE-1&news=4F2292E9-3EFF-48D3-A7E4-CEFA05D70C21>

'Tuvalu threat,' ABC Local Radio, Australia, AM Archive, transcript from 4 March 2002. Available from: <www.abc.net.au/am/stories/s495507.htm>

Viewpoint of the Federated States of Micronesia on the complex renovation of Prunéřov II power plant 3x250 MWe plan. Letter of 4 January 2010 from the Director of the Office of Environment and Emergency Management of the Federated States of Micronesia to the Ministry of the Environment of the Czech Republic. Available from: <http://aktualne.centrum.cz/domaci/fotogalerie/2010/01/06/dokument-dopis-mikronesie-k-planu-na-rekonstrukci-/foto/287153/?cid=657469>

Selected Internet Pages

<http://climatechange.worldbank.org/climatechange/>
<http://epa.gov/climatechange/EPAactivities/regulatory-initiatives.html>
<http://unfccc.int/essential_background/convention/status_of_ratification/items/2631.php>
<http://unfccc.int/home/items/5264.php>
<http://unfccc.int/home/items/5265.php>
<http://unfccc.int/kyoto_protocol/registry_systems/itl/items/4065.php>
<http://unfccc.int/playground/items/5516.php>
<www.foe.org/pdf/OPIC_Settlement.pdf>
<www.icao.int/environmental-protection/Pages/climate-change.aspx>
<www.imo.org/OurWork/Environment/PollutionPrevention/AirPollution/Pages/GHG-Emissions.aspx>

Index